Cerebral Entanglements

Cerebral Entanglements

How the Brain Shapes Our Public
and Private Lives

Allan J. Hamilton

Post Hill
PRESS

A POST HILL PRESS BOOK
ISBN: 979-8-88845-948-5
ISBN (eBook): 979-8-88845-949-2

Cerebral Entanglements:
How the Brain Shapes Our Public and Private Lives
© 2025 by Allan J. Hamilton
All Rights Reserved

Cover design by Ariel Harari

Post Hill Press
New York • Nashville
posthillpress.com

Published in the United States of America
1 2 3 4 5 6 7 8 9 10

*To my wife, Jane,
and our children, Joshua, Luke, and Tessa,
for their endless support and love
through a carreer obsessed with the brain*

I am a part of all that I have met;
Yet all experience is an arch wherethro'
Gleams that untravell'd world whose margin fades
For ever and forever when I move.

—Alfred, Lord Tennyson, "Ulysses"

To the Reader

This book is intended for informational purposes only, and is not intended to be used for any medical application. It is not intended for the diagnosis or treatment of any illness or condition. Please be sure to consult your health-care provider about any questions or concerns that pertain to your own well-being or health. The opinions expressed in the book are the author's personal views.

Contents

Cerebral Entanglements

Introduction

There is no scientific study more vital to man than the study of his own brain. Our entire view of the universe depends on it.
—SIR FRANCIS CRICK, Nobel laureate

Breaking Into the Cranial Vault

As a brain surgeon, I have made my living for more than three decades by "breaking and entering" into patients' skulls, much like a bank robber. In fact, the skull is anatomically referred to as the cranial vault. As with any good bank heist, there are two important rules. First, know where the money is. Second, get in and out as fast as you can. No alarms. And no one gets hurt.

How and where you open the cranial vault is critical. There needs to be an *exact* correspondence between where you create the opening in the skull—the so-called craniotomy—and the anatomical structures beneath, to which you want to gain access. You need to know exactly where the money lies. And gaining access—getting through the skull—is no easy task. The human skull can withstand more than five hundred pounds of weight on it before cracking. That's like balancing two full-sized refrigerators on it. And its walls are remarkably dense. To gain entry into the skull, you need quite a drill. Not one you might pick up at the Home Depot. This industrial-strength, pneumatic drill is powered by hundreds of pounds of pressurized nitrogen, with the working end made up of an inner ring of fluted stainless-steel teeth that spins inside a larger outer drill bit with even sharper, larger teeth. It makes real holes, big enough that you could sink your thumb through them. And it can remove a quarter-sized plug of incredibly dense bone in two or three seconds. If your surgical glove gets accidentally snagged in it, it will take your hand off in about half that time.

When the power button is pressed, there's a momentary shudder as the twin bits jump to life, cutting through all three layers of bone that lie over each other like sedimentary rock. In the process, the tip generates so much heat from friction that it must be irrigated with a stream of cold saline to keep it from cooking the valuable contents beneath. The first time I saw a surgeon drilling holes into the skull, it reminded me more of an operation on an oil rig than a surgery suite. It had that feel of industrial might.

This massive drill is designed with a spring-loaded clutch that keeps constant downward pressure on the menacing bit until it feels a slight give in the bone, when there is just the final millimeter or two of bone left. It's ingenious: when the clutch senses that little elastic give, it is engineered to instantly fly back and disengage the drill from the spinning shaft. Only once in my life have I seen what happens when the clutch fails. It happened to a colleague of mine in the operating room next door. A blood-spattered nurse burst through the door, yelling for me to come help. I leaped into a new scrub suit and ran, but the brain had been so badly ripped apart it was sickening. That's why we test that clutch every time. Every single damn time.

After a few of these preliminary holes—burr holes—are drilled into the skull, we cut the bone in between them with a side-cutting instrument, spinning at sixty thousand rotations per minute. This one has no clutch. When the saw has cut into every burr hole—when you've connected the dots—then, as we say, you've "cracked the coconut." All that's left is to peel back the skull flap, cut through one thin membrane, the dura, and you're in.

And there it is: the brain. It has no pain receptors of its own. Once we've drilled, sawed, and blasted our way in through the skull, we are inside what I call the Temple. The holy of holies. The brain sits there defenseless like pulsating, vascularized yellow custard. We can't even really cut it. We have to suck our way through it. Granted, we do it with small handheld suckers, machined by compulsive engineers down to the micron. But they are, even in medical jargon, still called suckers, and the sound they make is like a kid cleaning out the bottom of a milkshake, only they are sucking out brain tissue. The drab, underwhelming appearance of the brain is disarming: what a simple looking vessel for the most wondrous and complex object in the universe!

As neurosurgeons, we are all universally humbled by the brain. We struggle to grasp the grandeur and scale of its cytoarchitecture. Every day we confront its unforgiving fragility—be it clinical or just plain physical. How does one coherently understand an organ that bears an uncanny resemblance to an overturned bowlful of curdled custard and yet embraces as many interconnections as there are elementary particles in the universe? It seems as though the brain must be another of these divine riddles, like the Holy Grail or quantum entanglement. Deceptively simple packaging but, oh, what an unbelievably imaginative, mind-blowing, cosmic creation. It holds an entire being in its connectome. Granted, if you found it in your driveway, you might feel inclined to scoop it up with a dustpan because of its outward appearance, but once you become aware of its true nature, you might find yourself wanting to kneel before it. Because it is the embodiment of the truest mystery: it defines a human being.

As brain surgeons, we spend half a lifetime honing our manual skills so that we learn to make the touch of our fingers gentler than a breath. I have likened good brain surgery to trying to sneak up and kiss a rattle-snake on the head without waking it up. We have to pull it off because we are maneuvering through a world of intricate magic, where the slightest inadvertent touch can bring doom, everlasting damage.

It may just be the appeal of the brain's extreme fragility and daunting complexity underscored by its underwhelming outward appearance, but sooner or later every neuroscientist risks falling prey to what can only be called a kind of neuro-mysticism. The transcendent spell of the brain. I believe if there is one thing that might make you believe in God, it is looking at the human brain. The beauty of the simple chalice and what it holds is breathtaking.

Cerebral Entanglements

In October 1927 twenty-nine of the most notable physicists met for an international meeting in Brussels, Belgium. It included some of the greatest geniuses humanity has ever assembled. The attendees included Albert Einstein, Niels Bohr, Marie Curie, Erwin Schrödinger, John von Neumann, and Werner Heisenberg. More than half of the individuals in attendance would go on later to be awarded their own Nobel Prizes! During the conference, the attendees struggled with a thought experiment related to what is called quantum entanglement. Entanglement predicts that if one measures the properties of a particle and maps its location in one part of the universe, then that very act will instantly define and bring into existence an identical particle elsewhere in the universe—even if those two particles are at opposite ends of the known universe. So, while any information about the properties of one particle would take billions of years—even traveling at the speed of light—to dictate those of its twin counterpart on the far side of the universe, they both simply come into existence with the same characteristics. For quantum entanglement to work, to have one particle instantly bring its twin into existence, meant that time and space would have to "collapse" to hold true. It seemed an absurd notion, and it was for this reason that Einstein utterly rejected the hypothesis, calling it nothing more than "spooky action at a distance."

It took nearly a century and three generations of physicists to carry out the thought experiment first imagined in 1927. In November 2016, hundreds of astrophysicists, working in more than a dozen laboratories on five continents, used two very bright stars (known as pulsars) that were millions of light-years apart to prove that quantum entanglement exists. In fact, it is a fact. And collapse or no collapse, quantum physics does accurately describe and predict some very unusual properties of our universe.

Seeing Human Thought

As a brain surgeon and a neuroscientist, when I heard about quantum entanglement, it leaped out at me as the perfect metaphor to explain how we have come to understand the true *meaning* of the human brain in the twenty-first century. If we look at properties of sensory perception, we see that they are properties of our own thoughts and awareness. They are properties derived from our brains, from our inner space. On the other hand, those perceptions correspond to a set of physical or sensory properties that we detected in outer space—that is, the physical world around us.

Let me use a simple thought experiment to make the point. Imagine I shine a flashlight in your eye. The stream of photons from the light will impinge on the retinal cells in your eye. This will, in turn, engender a predetermined set of neuronal interactions in your occipital cortex and you will "see" the light. On the other hand, if I take that same configuration of brain cells and I simply stimulate them electrically, you will have the identical experience of seeing the light—even if the flashlight stayed in my pocket. While our perception of the outer world is dependent on the stimulation of our brain cells, the activation of our neuronal networks is what sustains the outer, physical world. These two worlds are created synergistically. They are caught up in a kind of cerebral entanglement: one mutually defines the other. They bring each other into existence.

Our personal lives depend on this neuro-physical communion, and our being—our identity—flows from it. The physical world determines the properties of the brain's activities, and those operations, in turn, define our perception of the physical world. It is a never-ending cycle of perceiving and being perceived. Our inner world of experience echoes the outer worlds of perception. The existence of one is intertwined with the existence of the other. In the past, only our outer world was accessible to us, but now, with the latest brain imaging (and its supporting neuroscience), we have been given the key to the equivalent inner one.

The overarching premise of this book is this: we are the first generation to be able to image and quantify human thought. To be able to put an emotion into pixels of activity on a scan. We live in a new age of insight in neuroscience, where we can analyze how our thoughts and emotions arise and what parts of our brains empower them. That analysis alters our understanding of ourselves and each other. It gives us a new perspective on how the collective actions of conscious and subconscious systems in our brains shape the lives we lead in private and in public.

Heady stuff, I know. Let me give you a simple but telling example of how neuroimaging fundamentally changes how we view the world. Until a decade ago, if we wanted to know if someone was telling the truth, we might resort to a polygraph test (a "lie detector") to try to assess if an individual was answering us honestly. Polygraph readings, however, can be notoriously unreliable. Now imagine this person is on the witness stand, lying under oath. He tells the jury that he *saw* the defendant commit the crime. The witness, who made up the whole eyewitness account, now has a powerful hold over what that jury may decide.

But let's add a brain imaging study (like a functional magnetic resonance image, or fMRI) into the equation. With the fMRI, we can determine if our witness *actually saw* the crime. Of course, we would have to have the witness lie down inside a large brain scanner as the defendant's attorney was posing his or her questions, but let's put logistics aside for a moment. Neuroimaging changes how we can understand the nature of the truth. If the witness is lying, his brain scans will show that his answers begin in the frontal lobe. That is because he has to *think* hard about how he is telling a lie, about how to fabricate his story. But if he actually did *see* the crime being committed, then his answers will originate in the occipital lobe, where vision is processed. *This* is what I mean when I say we live in a new age where functional imaging changes how we understand the meaning our brains give to our lives.

The Man at the Center of the Ring

When I was thirteen years old, I saw a news story about a doctor who claimed he had a new surgical technique that could—literally—stop a wild animal in its tracks. As I watched, I saw him stop a charging bull with just the push of a button. And that moment altered the course of my life. It is one of those things that still resonates with me to this very day, more than half a century later. Why? Because, at that moment, to me, he looked like the closest thing to a true wizard I could possibly imagine. And all I could think about was this: I wanted to acquire the same magical powers he did.

The name of this real-life Merlin was Dr. José Manuel Rodriguez Delgado. And I came across this flesh-and-blood sorcerer standing in front of the TV cameras at the center of a bullring in Ronda, Spain. He seemed to be standing there alone at the center of the world. But for you to understand what that spot meant, I must take you back half a century before Delgado was born.

It began with a bullfighter named Juan Belmonte. For centuries, matadors had been trained to use their prodigious speed to jump sideways to get clear from the path of the charging bulls in the ring. They would hold their muletas, small red capes, in the air as the bull charged forward. The animal would lower his head, trying to sink his horns into his human opponent, while the matador's artful, flashing movements of the muleta urged the bull to press his attack. Finally, the bull would hurl himself at the muleta as it passed over his eyes. In that flash, the matador would jump backward out of the reach of the bull's horns.

As a child, growing up in Triana, Spain, Juan Belmonte had dreamed of becoming a bullfighter, but the odds were stacked against him. He was unable to do what other matadors could do because he had been born with significant deformities; his legs were misshapen and atrophied. For Belmonte, having to jump suddenly and swiftly out of the bull's path was out of the question. If Belmonte wanted to fulfill his life's dream, he would have to come up with a different strategy to confront the bull.

Just how radical Belmonte's approach would be, no one could have guessed. In 1910, Belmonte emerged for his first bullfight in that very same Plaza de Toros in Ronda, Spain. Like every other matador, he walked to the center of the ring and began to wave his muleta. But after that, nothing was the same. The bull began his charge forward. But Belmonte did not jump out of the way. He did not move a muscle. He was like a statue, intent on holding his ground, even if it meant the bull would annihilate him. Spectators gasped and jumped to their feet, holding their hands over their eyes to keep from seeing the horror of what they thought would unfold.

As the bull hurtled toward the diminutive Belmonte, his muleta seemed to collapse in midair. It precipitously furled itself tightly around his shoulders and back. The bull charged right at the cape, but as he turned to hook it with his horns, he was violently spun around Belmonte's trunk, and somehow—unbelievably—the bull missed him by inches!

Rather than employing his cape to divert the bull's attention away from himself, Belmonte used it to draw the bull in even closer, making him commit to plunging forward with his horns as he chased after the cape. The result was Belmonte was able to maneuver the bull around his body in taut circles. As the bull pawed and attacked, Belmonte never allowed his feet to move. Instead, he used his skills with the cape to draw the animal nearer.

Just when the crowd thought it could bear no more, Belmonte raised the stakes even higher: he sank to his knees in the middle of the arena. He made it clear he would forsake any chance of escape. There would be no way to save himself if he miscalculated with the bull. Then Belmonte teased the bull closer. Impossibly close. Indeed, with some passes, the bull would turn so tightly around Belmonte that the tips of his horns would scrape some of the seed pearls that were embroidered into the ornate fabric of Belmonte's matador suit. He did not stop that day in Ronda until every person was on their feet, standing on the benches, madly chanting his name until he finally delivered them by dispatching the bull.

On that day in Ronda, Belmonte forever changed bullfighting. By the time he retired, more than five hundred of his fellow matadors had died trying to imitate his techniques. No bullfighter has ever come close to him. Thereafter, Belmonte became known simply as El Pasmo—"the Wonder."

Like so many young men in Spain, José Delgado had been inspired by Belmonte's almost superhuman feats of courage. As a teenager, he himself had watched as bull after bull would sweep past the great matador in the ring in Ronda. It was for this very reason that Delgado selected the bullring there. It was no accident that Delgado returned to his hometown of Ronda. Like Juan Belmonte, he had a point to make, and he would make it in front of the world—something just as amazing as what El Pasmo had done. Let them gasp and scream for Delgado as they had for Belmonte! On May 14, 1964, Dr. Delgado stepped out on the bright sand of Ronda's bullring to face a famous fighting bull, named Lucero. This bull, however, had a unique characteristic: a recently acquired, freshly shaved, little bald spot on the top of his head. The bald spot on his scalp was where Delgado had performed a surgical procedure to implant an electrode and a small chip deep into the center of the animal's brain. The chip had two components—a transmitter and a receiver—and the wire lay right at the heart of a pea-sized area of the brain called the amygdala.

Delgado was no bullfighter. No resplendent *chaquetilla* for him. Instead, he wore a sweater and tie like a college professor. No muleta. Instead, he held a small remote-control unit, like the ones used to fly model airplanes. And no sword to deliver the coup de grâce. Instead, he had pulled out a long telescoping antenna on the remote. Because Delgado was a neurosurgeon, he had carried out extensive research in the laboratory on the amygdala. He believed this almond-shaped structure (amygdala means "almond" in Greek) in the temporal lobe was the source of violent, aggressive behavior. He had developed a technique that he asserted could stop the amygdala from working. He had published several papers on the technique in scientific journals, but hardly anyone believed him. Now, like Belmonte, he had decided he would make his case in public, in front of a worldwide audience, and prove he was right or perish in Ronda. Like Belmonte, he crossed himself as he entered the ring and then nodded to the stockman at the far end of the arena to lift the latch on the large wooden gates there. Once they swung wide open, an enormous bull appeared.

Lucero came out of the gate. He wheeled around a few times, as fighting bulls often do when they first burst into the arena. Then one of the matadors stepped forward and waved a cape in the bull's direction. Delgado's transmitter received a signal from Lucero's amygdala, and then he charged as any Spanish fighting bull is bred to do. This was a critical step, because it demonstrated that Lucero still exhibited all of the usual aggressive behaviors one would expect in a bull. Delgado's surgery had not interfered or blunted the animal's aggressive drive to charge. Every time Lucero lunged at a matador, Delgado would see the recognizable burst of electrical activity in the amygdala as a signal on his handheld device. After a few repetitions to make the point, the matador finally folded his cape and stepped away. Now Delgado stepped into position. Wearing spectacles, with his tie under his pullover, he looked very much like a doctor about to make a house call. And he was armed only with the handheld transmitter a bit larger than a paperback book, with its long retractable antenna.

Delgado knew Lucero had spotted him. The bull snorted in his direction and began pawing the ground. Delgado saw the signal coming out of Lucero's amygdala. Then the bull charged. As the amygdala kept firing and the bull bore down on him, Delgado waited until the last minute and then pressed the button on his unit. It fired a series of electrical impulses back into Lucero's amygdala. The bull stopped dead in his tracks. He looked casually at Delgado, turned, and then walked off placidly as if nothing had happened. Delgado asked to repeat the experiment. The professional matador returned with his cape and rekindled Lucero's rage, and then Delgado stepped out and took over, waiting again until the last second before he pushed his transmitter button and brought the bull to a screeching halt. Delgado was like a new Prometheus stealing fire from the gods. Whatever emotional motivation was at the heart of the animal's violent reaction seemed to simply vanish when Delgado sent his blocking signal. Had Delgado tamed the darkest violent impulses with his wand?

Delgado was convinced he had discovered a fundamental principle of how the brain worked: impulses in one discrete area of the brain could set off specific, complex, and, in this case, violently aggressive behaviors. While working on his thesis, Delgado had been struck by the fact that the brain seemed engaged in an electrical conversation with itself, sending brain waves out from one area of the central nervous system (CNS), like a radio transmitter, to affect another area that served as a receiver. The receiving area would then send its own signal back. Delgado wondered if he could eavesdrop on the relatively small electrical signals given off by the brain. He implanted slender, hair-thin platinum electrodes into the brains of, first, cats and, later, monkeys. He found that he could detect when the electrical signals were being sent by the amygdala. If he sent a signal back, just as the neurons in the amygdala were starting to transmit, he could block them from working—almost like jamming someone's radar. In an instant, the amygdala (which we now know is responsible for generating rage) could motivate a one-thousand-pound bull with a desire to utterly destroy the human being in front of him. And in another instant, those impulses could be blocked and shut down.

Delgado's demonstration captured headlines. The idea that a scientist could take a dangerous animal and, quite literally, make it stop with an electric impulse suggested a new algorithm for understanding and changing the nature of the mind. I was captivated by Delgado's work for years. A few years later he wrote a book titled *Physical Control of the*

Dr. José Delgado stops a fighting bull during a full charge with a transmitter that sends a radio signal to the animal's brain to shut down its aggressive impulses.

Mind: Toward a Psychocivilized Society. In it, he presented a vision of a kind of utopian society of the future where mental illness and seizures would be amenable to control by precisely accessing and modifying tiny areas of the brain. It was a dramatic new take on human behavior: we change our minds by changing our brain function. Delgado's demonstration was a game changer. For many, it begged the fundamental question: Who's in charge? Do our brains run us or do we run them?

I contend that we have created an artificial dichotomy between mind and being because they are, in fact, completely entangled. As I said earlier: define one and the other instantly springs into existence. It is time, therefore, to write a new kind of book about the brain. Because we are the first generation to see images of a human thought or feeling. Today's neuroimaging gives us a brand-new lexicon for discussing how we emotionally and cognitively engage and negotiate the world around us. As physically insubstantial as thought or emotion may seem, we can actually see them take on a solid shape and form in the age of the digitally imaged brain.

We have powerful new technologies to evaluate the brain noninvasively. We can now peer into the CNS in real time. We can assess its function under normal, healthy conditions and see what goes awry in mental illness or behavioral disorders. Brain imaging allows us to compare one subject's brain activity before and after pharmacological intervention— something that is proving critical to refining treatment for mental disorders. We can also image the brain repeatedly so we can see how the CNS changes our emotional or behavioral responses over the course of development. We can compare individuals in their twenties with others in their seventies. For the first time, for example, we can answer the question, why, as teenagers, are we willing to engage in risky, dangerous behaviors that, later in life, as mature adults, we would never even consider for a second?

You might think the answer is attributable to something like increased experience or wisdom as we mature. But imaging technology now allows us to *see* wisdom's shape and form. I can now point to a particular set of structures, of fiber tracts, where insight, risk assessment, and self-restraint generated in the frontal lobe exert their inhibitory effect on more impetuous emotional centers in the brain. Another way to consider this is that, hypothetically, if one could surgically sever those bundles of fibers, you would, in fact, start thinking and acting like a teenager again. God forbid. But we do, see such behaviors when the uncinate fasciculus is badly damaged or disrupted with a severe head injury.

Finally, as has happened in so many other imaging methods within the human body, we are beginning to see the power of applying artificial intelligence (AI) to analyze regional brain activity in these images. This is an important step because a computer can acquire in a few hours the cumulative memory of many human life spans. No one human being could assemble the experience (let alone the recollection) from evaluating, say, a quarter of a million scans. But a computer can. AI has already proven itself more accurate in reading mammograms, electrocardiograms, biopsies, and retinal scans than human specialists. We can expect that AI is likely to hit many more home runs in the future of neuroimaging.

Three-Tiered Approach: The Brain, My Life, and Our Lives

We are obviously talking about a lot of new methods and information about the brain. Sometimes, it is complicated to extract the significance of what a new scientific finding is telling us. This book is therefore organized in a three-tiered approach:

1. The first tier is to explore how the latest neuroscience and imaging studies help us to better understand many of the important themes in human experience.
2. The second tier is to evaluate how these thematic insights apply to us in our personal, daily lives.
3. Finally, the third tier looks at how these insights shed light on the broader context of how we behave and react as members of society.

To this end, the book is divided into chapters that explore how the brain shapes emotional or behavioral themes like love, evil, happiness, and memory. To some extent, successive chapters will rely on some knowledge about the anatomy or function of a particular brain region that has been spelled out in an earlier chapter. However, it is my intention that readers can simply thumb through the listing of topics and turn directly to those that are of interest. *Cerebral Entanglements* is not intended to be a "self-help" book per se. Naturally, parts of it can help us to see ways to improve ourselves. To reimagine ourselves. Or to forgive ourselves. This book is intended to provide a map of the major cognitive and emotional destinations in life and provide us with a greater appreciation of how they are woven into the landscape of our personal lives and society at large. Ultimately, what we do with such a map is up to each of us.

Throughout the book I have added my own insights and experiences about brain function in my capacity as a neurosurgeon. I have been able to devote a gratifying and satisfying lifetime to studying the brain precisely because I am still learning so much more about it. I am still enamored with it and awed by it. And it still surprises me every single day.

Imagine for a moment that you are an Indiana Jones-type character. As your guide, I owe it to you to explain that when we push against the mysteries of the brain, it will take us to other far-flung places. We are truly inside the temple, the skull, confronted with the vivid, vital jewel-like essence of the brain, shimmering before us. But there's always that mystery lying ahead; how can we explain everything that's *out there* with everything that is *in here*? Together we will ask the tough questions: What is love? Laughter? Kindness? Music? Murder? Happiness? That's is why we are here: to find answers. But like Indiana Jones, we finally pry open a door leading to a secret passageway. We peer in together. Now, we see a long way ahead of us as we light our way, armed with new scans, functional imaging, and new experimental paradigms. And now we can train artificial intelligence to divine human kind.

I see two mysteries lying ahead of us The first is to explore the human brain so we may better apprehend our lives—private and mutual—together. We share so many themes in common, and yet in isolation. There may be some similarity to the mechanisms underlying grief, for example, but the expression of that grief is entirely unique to each of us. The trick will be to understand the shared parts of the individual thoughts and emotions we express. The second mystery is cosmic in nature. The ultimate pursuit of the human brain is to grasp the size and scope of the universe. Here on Earth —and maybe in myriad other interstellar birthplaces—our intelligence is evolving toward cosmic insight and understanding. I suspect that the journey into the nature of our own brain is what will carry us to the heart of creation. We are truly made of matter from the stars themselves. And inside of each of us twinkles a universe of neurons, shining with the same light and same creative power that embraces the cosmos. We peer into ourselves at the same time we gaze into the depths of space. We are two twin universes, married, I suspect, to a greater design than we ever imagined. Let's push on, shall we?

1

On the Nature of Consciousness

Wish I didn't know now what I didn't know then.
—BOB SEGER, "Against the Wind"

Consciousness as Being

In 1964, the Supreme Court tried to issue a ruling on where art ends and pornography begins. In writing his opinion, however, Justice Potter Stewart could not provide a clear definition. Instead, when it came to pornography, he surmised, "I know it when I see it." In that regard, consciousness is like pornography: hard to define, but we recognize it when we see it.

Despite that recognition, the nature of consciousness is elusive. Sir Francis Crick was a luminary in the scientific world for his role discovering deoxyribonucleic acid, for which he shared in the 1962 Nobel Prize. Despite his enormous achievements in molecular biology, Crick felt that consciousness was the most critical issue that science had to address. He dedicated the last half of his life to its pursuit. In his research on the linkage between self-awareness and consciousness, he collaborated with Christof Koch, a scientist at the Allen Institute for Brain Science. Together, they spent a quarter century trying to chase down what brain structures brought about and sustained our state of self-awareness.

Crick wrestled with consciousness right up to the end of his life. He was working on completing a manuscript about consciousness he had written with Koch when he passed away. He ended up describing consciousness as "'you,' your joys, and your sorrows, your memories and your ambitions, your sense of personal identity and free will." His depiction sounds almost poetic because consciousness is one of life's essential ingredients while remaining one of its most mysterious.

Consciousness has two meanings in neuroscience. The first is self-awareness. That was what Crick and Koch were obsessed with working out. The second relates to the brain's ability to recognize sensory input from the body. We have to address both because they are two sides of the same coin.

Consciousness as Self-Awareness

Consciousness is the sense that each of us has a unique identity. That we are a being. A presence. Two thought experiments will illustrate this notion from different perspectives. For the first, imagine you are given twin babies born with identical genomes at birth. And you raise them under the most rigorous experimental circumstances you could ever design. They would eat the same food. Attend the same classes. Wear the same clothes. Have the same friends. No matter how much you attend to making their environment the same, you will always end up with two different beings. A separate consciousness emerges in each twin as they grow. One twin will never mistake him- or herself for the other one. There is no confusion between them as to who is twin A or twin B. This thought experiment shows that each person develops a unique existential insight into his or her own being.

The second hypothetical experiment involves a patient who might be lying close to death on the operating table. He has an out-of-body experience: the patient suddenly sees himself rising in the air and looking down on the scene as the surgeons work to save his life. This is a scenario that a small subgroup of patients has experienced in well-documented cases of out-of-body experiences. So, I ask you, in this out-of-body experience, who is looking down on the OR table? Who is perceiving the scene? The answer is the patient's consciousness.

We are looking at an experiment routinely used to test for self-awareness. You should imagine a simple scenario where a dolphin looks into the mirror and says to itself, "Oh, look, another dolphin is staring back at me." But the dolphin has received a small, temporary stripe applied to the top of its head with a harmless dye. It now becomes an entirely different matter if the dolphin swims over to the mirror and then proceeds to look at itself and ask, "Hey, what's that mark on the top of my head?"

This self-awareness, or what the neuroscientist Antonio Damasio terms the "autobiographical self," is no trivial matter. In the entire animal kingdom, only a handful of species possess minds complex enough (dolphins being one of them) to pull off this neuronal "magic trick." Human babies only learn to recognize themselves in the mirror when they are eighteen to twenty-four months old. I might add they seem quite pleased with themselves when they do! Self-recognition requires a vast network of synapses to create and sustain the emergence of an integrated identity. We know there are substantial, discrete areas of the human brain (much derived from Koch and Crick's work) devoted to helping us recognize images of ourselves and these are distinct from other loci where we recognize others.

With the latest brain imaging technology, we can ask, what happens in our brain when we recognize ourselves? Something that we take for granted when we see ourselves in the mirror each morning. When we

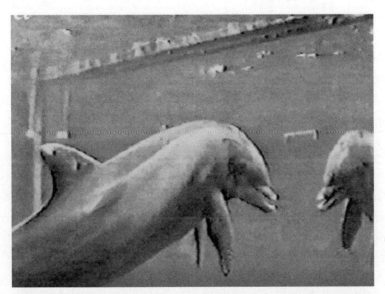

A dolphin looks at his reflection to assess a temporary mark applied to the top of his head. The "mirror test" is a litmus test of self-awareness.

take a group of normal subjects and have them hop onto a functional MRI scanner, we can flash a series of facial photographs on a computer screen and see what happens when we randomly insert photos of the subjects themselves into the series. With the fMRI, we can measure the level of brain activity in a specific area, the *fusiform gyrus,* where the subject recognizes him- or herself. But now watch as this simple act of self-recognition blossoms into a more complex insight. If we do the same experiment but this time we take a group of subjects who have been diagnosed with moderate to severe depression, we will see far less activation of the fusiform gyrus on their brain scans. When computers use AI to quantify the amount of activity in the fusiform gyrus, they can reliably predict if the fMRI images belong to an individual from the normal or depressed cohort.

We know that depressed individuals often suffer from lower self-esteem. But now we see the neural correlate of it in the diminished activity in the fusiform gyrus: literally—a decrease in self-recognition. Andrew Solomon wrote about this lack of self-esteem in his beautiful book, *The Noonday Demon: An Atlas of Depression:*

> When . . . [depression] comes, it degrades one's self and ultimately eclipses the capacity to give or receive affection. It is the aloneness within us made manifest, and it destroys not only connection to others but also the ability to be peacefully alone with oneself.

Neuroimaging is still a kind of metaphor. We can now see and measure how the different parts of the brain react to a stimulus. And, yes, it is a giant step forward in understanding the interactivity between a stimulus and the CNS. While we can add up the intensities of pixels (tiny dots of light) on an fMRI display screen, we are still learning to decipher what these displays mean. We can state that self-recognition arises in the fusiform gyrus, but we can only postulate why diminished activity is associated with depression. And we cannot say which came first or why. That self-recognition is partitioned to its own locus underscores the significance the brain assigns to the value of self-recognition, but we are still struggling to understand how our mental health and self-image may be intertwined.

Consciousness as Brain Function

The second neuroscientific meaning of consciousness is a measure of the sensory responsiveness of our CNS. For this, think of consciousness as the ocean. When consciousness is first impaired, we sink just below its surface. Call it snorkeling depth. Light flashes everywhere, and there's a frolic of life-forms, of reefs and tides, plankton, and schools of fish, and we can also easily bob back up to the surface. This depth would be analogous to sleeping. We're easily aroused by a loud noise. As we sink deeper into unconsciousness, we gradually lose contact with the surface and, eventually, even sunlight. We can no longer quickly bob up to the surface now. There's far less sea life (that is, sensory activity) at this depth (see figure 1). As we sink lower, we finally plunge into black featureless fathoms. Here, there is no more up or down. There is no sea life. Here, we find ourselves in the most profound depths of unconsciousness, in the realm of coma and brain death. Life at these empty depths is stripped of being.

A Quick Primer on Brain Imaging

Throughout the book, we will return to brain imaging studies to provide us with insights about the brain. There are several different types of brain imaging studies that will be referenced. They use different methods to get back to one central purpose: to find out what regions of the brain increase or decrease their activity in response to what the individual is experiencing. The methods all work by a basic process known as subtraction.

We begin subtraction by capturing a series of image slices of the brain at rest when it is not doing or thinking about anything. Even at rest, however, there are lots of things happening in the brain. The person is breathing, seeing, even just feeling the scanner around the head. But whatever images we obtain under these conditions serve as our baseline control (see figure 2). Next, we obtain another set of images while the brain is "stimulated." In other words, we can compare brain activity while it is responding to a stimulus (for example, a photo, a recollection, or an emotion). This could be an activity as specific as when a test subject is feeling anxious, lying to a police detective, or just petting their dog.

We measure the amount of brain activity seen when the brain is stimulated and subtract everything that was on the baseline control images. Once we have subtracted all the background baseline activity, we will see only that regional brain activity that was specific to the stimulus we introduced. We can use the same subtraction technique for all three basic types of brain imaging studies, or scans, at our disposal, namely, functional magnetic resonance imaging, positron-emission tomography (PET), and single-photon emission computed tomography (SPECT).

Functional MRI is, by far, the most popular of the modalities. It looks at minute changes in blood flow and oxygen delivery as brain cells are activated. PET scans use radioactively labeled compounds (like glucose) given intravenously to directly image how active the neurons are by gauging how much energy they consume. SPECT is the least commonly used technique, and it measures radioactivity signals emitted by an administered isotope to give us 3-D images of brain activity. The common feature of all three is we can measure momentary fluctuations and compare them with either an individual's baseline activity or a group average at rest.

Consciousness: Coma, Persistent Vegetative States, and the Hope of Brain Imaging

"Loss of consciousness," "coma," and "brain death" may be everyday terms with which we all feel familiar because they have been around for a long time in medicine. As far back as Hippocrates (460 BCE—375 BCE) and Galen (AD 129—ca. 210), coma (derived from the Greek *koma*) meant "a deep sleep" and was recognized as being related to brain injury. In the neurosciences, different levels of consciousness came to be recognized as "altered mental status," the assumption being that anything less than sharp, crisp mentation represented an abnormal condition.

Sometimes, the right patient comes along serendipitously. And that is what happened on December 14, 1650, when Anne Greene, a scullery maid, would make an unknown Oxford physician named Thomas Willis instantly famous. So famous, in fact, that he would be celebrated by the British monarchy and eventually buried in Westminster Abbey.

Anne Greene was an uneducated young woman who was working in the household kitchen of Sir Thomas Read, a magistrate justice in Oxfordshire, England. According to court records, Greene claimed that she had been seduced and impregnated by Sir Thomas's seventeen-year-old grandson. In court, she testified she was unaware of becoming pregnant from the liaison until she miscarried more than four months later. She buried the remains of the fetus, but they were later discovered. Greene was charged with not just murder but infanticide. Her prosecution was conducted by none other than Sir Thomas himself, her former employer. She was found guilty of murder and hanged on December 14, 1650, in Oxford Castle.

At that time, public hangings were grisly affairs, with the executed struggling and squirming for many agonizing minutes after the noose was pulled taut. Most executions used "the short drop method" for hanging. Think of the typical lynching one sees depicted in Westerns where the cowboy dangles from a noose after his horse is whipped out from under him. Or when individuals hang themselves by standing on a chair and then knocking it out from underneath their feet. There is not much of an actual drop powerful enough to produce a quick snap of the cervical spine. Instead, the victim suffocates from the noose slowly strangling them.

Anne Greene was well aware of this, so she specifically enlisted several friends and assigned them the task of yanking down very hard on her body by the feet to ensure she had a swift demise. She confessed to the crime on the gallows and was hanged. As promised, her friends jerked forcefully down on her swinging body by the ankles several times. More than half an hour later, a soldier hit her three or four times with the end of his musket to make sure she was dead, then cut down her body and placed it in a coffin. As was the custom of the day with any executed criminals, attendants delivered Anne Greene's body to the nearby anatomy department at Oxford University, where she and a young physician named Thomas Willis were about to make medical history.

In the mid-seventeenth century, timely dissection of the fresh body was the rule because there was no means of preserving the corpse. The first thing the following day, Willis and his assistant opened her coffin and laid the body out on the dissection table. As they examined the injuries to the neck caused by the noose, Willis detected a very faint pulse in the carotid artery. A wide and wild assortment of remedies were immediately brought into play in a desperate attempt to revive her. It is a testament to Anne Greene's fortitude that she survived both the hanging and the subsequent therapy used to revive her.

First, Willis tried pouring alcoholic spirits down her throat. After that, he vigorously massaged her limbs. Then he commenced bloodletting, followed by the application of warm poultices to her chest. Probably the most bewildering therapy Willis employed, however, was the insufflation of the colon with warm tobacco smoke delivered via a bellows placed in the rectum in the hopes of warming up her internal organs. After that, Greene was placed in a warm bed alongside another woman to ensure the application of body heat. I imagine this must have been one of those night-shift assignments that got doled out to the individual with the least seniority.

By the following day, Anne Greene was able to speak. Her throat was badly injured, but she was able to eat solid food in another four days. She recovered fully in a month and was mercifully completely amnestic about events surrounding her hanging (and the therapy). In a strange twist of fate, Sir Thomas Read died suddenly three days after her execution. The whole business became wildly sensationalized. The publicity surrounding the astounding revival catapulted Willis to fame. Greene was eventually pardoned, because events were believed to convincingly demonstrate that divine intervention had played a role in her survival. She returned to the home of a family friend six weeks after her execution. She insisted that the coffin accompany her on the journey.

Later, she married and bore three children. Willis would go on to become the royal physician and one of the founders of the Royal Society of London. He coined the name "neurology" for the study of the nervous system and made scores of discoveries about brain anatomy and function.[*] He penned many scientific works, including a famous treatise in 1672 that he titled *De anima brutorum* (*On the Soul of Brutes*). His accumulated experiences with different brain injuries led Willis to classify mental status into several categories. He assigned the term "lethargy" (a term we still use today in medicine) to describe someone difficult to arouse. He correctly attributed the condition to altered function in

---

[*] The circle of blood vessels that supply the brain is named the circle of Willis after his description of the unusual circular blood flow that helps ensure a sustained flow of arterial, oxygenated blood.

the cortex, the dense layer of neurons covering the entire surface of the hemispheres. He designated the term *carus* (from the Greek root for "sleepy") to describe the condition where the individual has no sensory responses. He gave us the term "coma" for what he called heavy sleep, like the unresponsive state in which he found Anne Greene.

It was not until the early twentieth century that it was clear that a worsening state of consciousness was due, as Willis had suggested, to the progressive disruption of brain function in the cortex. A special kind of unconsciousness was labeled by a French neurologist *coma vigil* (which translates into "awake coma"). The term described a state where the individual was able to open his or her eyes, still had some brain-stem reflexes (like swallowing or coughing), but was utterly unconscious and unresponsive. Later, the condition was renamed in English a persistent vegetative state, referring to the patient's condition as "wakefulness without awareness."

The term "vegetative" was an unfortunate choice of words. It had been selected by the British scientists Bryan Jennett and Fred Plum because they referred the term back to the verb "vegetate." In defense of their word choice, they explained that the term was meant to convey the idea that patients in such a state were living "a merely physical life devoid of intellectual activity or social intercourse" and meant to describe "an organic body capable of growth and development but devoid of sensation and thought." The term has since been badly misinterpreted and led to the vernacular of patients being called "vegetables"—a term I detest.

For these reasons, the medical community has turned to using the expression "unresponsive wakefulness syndrome." However, you still hear "persistent vegetative state" a great deal in hospitals and the medical literature. And, unfortunately, many thousands of individuals a year persist in such a pitiful state—whatever name we give it. Some patients have survived in an unresponsive wakeful state for as long as three decades without a glimmer of hope of regaining higher cognitive function. Or so we thought until 2006.

That is when an impish, red-haired scientist named Adrian Owen was reading an fMRI scan at his workstation at Cambridge University that changed forever the way we think about coma. From an academic point of view, Owen is a kind of whirling dervish. He has produced dozens of scientific journal articles and books to persuade the medical world to rethink some of its fundamental notions about coma.

Back in 2006, while Owen was looking at fMRIs, Roger Federer was engaged in a furious finals match with Rafael Nadal at Wimbledon. He would defeat Nadal 6–0, 7–6, 6–7, 6–3 to win his fourth Wimbledon title. As the historic match gripped the world, Owen was in the process of evaluating a twenty-three-year-old woman in an unresponsive wakeful state as she lay on the table of an fMRI scanner. Listening to the match in the background—they can be deathly silent except for the thwacks back and forth and the calls from the judges—he halfheartedly asked this comatose young woman to imagine *she* was the one playing tennis. Remember that earlier I talked about how a brain scan is a study in subtraction. Owen had a baseline. The scanner had already captured images of her brain as she lay there comatose. He expected to see no response at all: baseline minus baseline equals zero. But to his utter surprise, he could see areas on her scan lighting up with activity. It appeared as if her brain were trying to direct her eye movements to track the volleying of a ball back and forth. He could also see motor activity in her cerebral cortex, as if she were preparing to swing and hit the ball back after she was tracking it!

The neuroimaging results would lead Owen (and his fellow researcher Steven Laureys) to begin hunting down dozens of patients in unresponsive wakefulness syndrome for evaluation. What they found surprised the whole world of neuroscience. Contrary to what neuroscience had been taught to believe—that almost all unresponsive wakeful patients were hopelessly doomed to stay in a coma for the rest of their days—a significant percentage of those languishing patients were trying to respond based on their fMRI results. While the world had given up on them—I heard one physician refer to them in a lecture as "living human paperweights"—many of them were trapped inside their brains, battling to wake up! Once these individuals could be identified with neuroimaging, aggressive testing and rehab could help salvage them and help to stimulate their brains to reawaken.

One such patient was a twenty-six-year-old named Kate Bainbridge. She had been a victim of a devastating brain infection that had left her in an unresponsive wakeful syndrome. Fortunately, her condition also qualified her for inclusion in a series of experimental studies conducted by Owen's research group at Cambridge. Comatose patients were shown a series of random photographic images. But interspersed among them were photos that would *only* be familiar to the patient. In Bainbridge's case, the evaluation revealed that her brain began recognizing photographs that were unique to her life as different from the rest of the photos. She could distinguish them accurately and reproducibly from the other random pictures. Eventually, Kate would regain consciousness and, ultimately, her language function. She can tell us in her own words about the ordeal that too many patients caught in unresponsive wakefulness syndrome face:

> Not being able to communicate was awful—I felt trapped inside my body. I had loads of questions, like "Where am I?," "Why am I here?," "What has happened?" But I could not ask anyone—I had to work it all out. I could not move my face, so I could not show people how scared I was.

Owen's neuroimaging studies in comatose patients represent a significant breakthrough that could only come about because we are the first generation of human beings to see human thought with brain imaging. Some estimates suggest that as much as 25 percent of patients in the United States deemed to be in a hopelessly unresponsive wakefulness syndrome may demonstrate reactive activity if they are evaluated serially with the latest brain imaging technology. These results have brought a worldwide urgency to harnessing brain scans to help us identify those patients who are trapped in a coma, waiting for us to find them, and help them escape. To bring them back to the surface and feel once again the mysterious, warm light of self-awareness.

Neuroimaging has given us a deeper insight into the nature of consciousness than we've had before. We now know the term "unconscious" refers to a condition where awareness is *unavailable* to the central nervous system. Think of it like a broken cell phone. It does not matter how much bandwidth or how many conversations are going on, because we will not hear any of them. While the "subconscious" refers to brain activity that is *excluded* from our awareness, we are listening to that fraction of the conversations occurring on the network designated for our cell phone. But there are millions of conversations going on elsewhere we cannot access.

Cognitive Bandwidth: The Conscious and the Subconscious

What the latest brain imaging technologies reveal about our subconscious is humbling. It brings us to the first tenet of the brain: namely, the vast majority of everything that is affecting us in the physical world is beyond our ability to know it. We understand now the brain's "cognitive bandwidth"—the fraction of brain activity we reserve for conscious perception, cognition, and emotional processing—is exceptionally narrow. About only 100 bits of information per second. To put this "brainpower" into context, when you and I are carrying on a conversation with each other, that requires about 40 bits a second. If someone else is in the room talking away within earshot (or the TV is going on in the background), that's another 40 bits. The last 20 bits gets eaten up by things like the clothes we're wearing, the texture of the seat cushion we are sitting on, and swatting at the fly that keeps buzzing around our head. That's it. You maxed out your conscious bandwidth.

We, therefore, have an extremely narrow window—a tiny trickle of information—through which we consciously perceive the world. What we are finding out is that the sensory traffic flowing into the brain, on the other hand, is enormous: on the order of 10–11 million bits per second! To illustrate the sheer scale of this discrepancy between conscious and subconscious input, if our conscious bandwidth were represented by the two hundred-mile-distance between Boston and New York City, then the subconscious traffic would generate a distance equivalent to circling the equator 275 times! Only 0.00005 percent of all the incoming traffic is being routed to our conscious thought processes. Everything else we are experiencing about the physical world through our five senses is diverted to subconscious pathways. This deluge of millions of sensory and autonomic signals is pushing up through the subconscious centers in the brain stem, cerebellum, and midbrain but will never see the daylight of our own conscious awareness. We can never glimpse the whole picture.

Now, however, we are able to ask, how do these currents—one a trickle and the other a tsunami—interact? The first approach is simple editing. The brain shunts the highest priority, most relevant data to our conscious awareness—the trickle. The rest is all on autopilot and gets shunted to lower centers of the brain, such as the brain stem, midbrain, and subcortical structures. So, if our pants catch fire, that goes straight up to the top. But it is not a one-way street, either.

Conscious processing frequently gets handed off to the subconscious. Take, for example, bicycle riding. When we first learned how to ride, maybe one of our parents ran nervously behind us. Or perhaps we had training wheels. We went through a clumsy stage where steering and pedaling simultaneously—and stopping—were challenging. We might have even run into one or two stationary objects. But now? We hop on that bike and take off. We do not even think about locomotion and steering. We lean into the turns to adjust for centrifugal force as simply as breathing. It has all been shoved over into the subconscious so our limited conscious bandwidth can be shifted to other things as we bicycle, like the terrain, the wildlife, or traffic around us. We no longer need to use a lot of conscious bandwidth for the routine tasks of bicycling, and they get shunted to subconscious handling. But what controls this assignment process? At this point, let me introduce you to the ascending reticular activating system—what, in neuroscience, could be called "the ghost in the machine."

The search for the ghost started with a jousting tournament held on June 30, 1559. We know the exact date because it involved Henry II, king of France, and the whole incident was painstakingly chronicled by eyewitnesses and historians of the day. On that date, the king was hosting a gala to celebrate the recent treaty ending the war between the French crown and the Habsburgs of Austria. It was also meant to mark the engagement of his daughter Elisabeth to King Philip II of Spain. This was a union that Henry had carefully cultivated through envoys to the Spanish court, who brought lavish gifts from King Henry II and exchanged hand-painted portraits of the betrothed. King Philip was a Habsburg, and the marriage to Henry's daughter was meant to wrest control of the Spanish court away from the Habsburgs and bind it instead to the French monarchy and its descendants. Henry really wanted to celebrate.

The king decided that the weekend's crowning event should be a lavish jousting tournament. And not just any tournament. The celebration would end with what could only be termed a headliner act. It would feature Gabriel, the Count of Montgomery, the dashing captain of the king's own Scottish Guards. And his opponent would be none other than King Henry II himself. As soon as the contest was announced, the captain tried to dissuade the king. The count's own family begged him not to proceed, but Henry was an avid jouster and could not be dissuaded from it. He wanted the whole celebration to end with a great spectacle. And it did. Just not the kind anyone expected.

As the two knights catapulted toward each other on their horses, the captain's lance struck King Henry squarely in the helmet, and then it splintered. The king reeled backward from the impact. Aides rushed forward to lower him off his horse. As soon as they did, they gasped. Part of the lance had penetrated right through the helmet and was lodged deep into the king's eye socket. He was bleeding profusely and was barely conscious and was slipping into a coma.

The queen immediately summoned the most famous and celebrated surgeon in France to be brought to court to personally attend to the king.

The jousting tournament where King Henry II was fatally wounded on the last day of June 1549.

His name was Ambroise Paré. Paré was legendary for some of the surgical innovations he had introduced into military surgery and the many astounding "saves" he had pulled off on the battlefield. He was rushed to the king's side and removed the piece of lance so he could inspect the wound. When Paré examined the length of the lance fragment, it was clear to him that it had penetrated deeply into the king's brain. Paré had decades of experience as an accomplished battlefield surgeon, and he knew the prognosis was dismal. The site soon became infected, and the king died a few days later.

Paré went on to coin two terms relating to the loss of consciousness resulting from head trauma. The first was "commotion" and the second was "concussion." He used commotion (*commotio cerebri*) to refer to situations where the brain has been badly shaken and subjected to great physical acceleration and deceleration, as had occurred in King Henry's case. Although we no longer use the expression "commotion," modern CT and MRI scans reveal the astuteness of Paré's hypothesis about motion inside the skull damaging the brain. "Concussion," of course, is still a term widely in use today, and Paré used it to describe an individual who received a blow to the head and fell unconscious but then later regained their full faculties. Paré was far ahead of his time in his medical thinking.

By the early twentieth century, two different schools of thought about consciousness had emerged. One group argued that unconscious states like coma resulted from the absence of cortical processing. The second school of thought argued that could not be the whole explanation. For example, normal individuals "lose consciousness" when they fall asleep but can still remember their dreams. Sleep was a scenario where unconsciousness and cortical processing were not mutually exclusive. The second group of scientists reasoned that there must be an additional caveat to consider: unconsciousness could also occur if sensory information were excluded from the cortex for processing.

The first school argued that the cortex had to be functioning for consciousness to emerge. Think of the cortex as a radio on the nightstand: the scientists in the first camp argued that there would be no music to be heard if the radio was broken. The second group argued that, yes, that would be one condition to consider. But the radio could also be in perfect working condition and turned on, but if the radio weren't properly dialed to the frequency of a particular radio station, music would still not be heard. The search was on for a structure in the brain that could determine when the cortex was or was not dialed in to receive the sensory signals being broadcast to it.

Santiago Ramón y Cajal was a gifted artist and anatomist from Spain. He became one of the foremost microscopic neuroanatomists of the early twentieth century and was awarded the Nobel Prize in 1906 for his descriptions of networks between brain cells. As part of his work on the brain stem, he was the first to describe a diffuse complex of hundreds of microscopic nuclei (clusters of neurons) that were embedded throughout the length of the brain stem amid an extensive network of fiber bundles. He labeled this new structure the reticular formation (from the Latin *reticulum* for "web"). The structure was so diffusely dispersed along the whole length of the brain stem that he was at a loss to ascribe any function to it. Only decades later, when scientists could accurately insert electrodes deep into the network, did they began to discover what Ramón y Cajal's reticular formation did.

The studies involve placing electrodes into the nuclei of the reticular formation of animals while they are fully anesthetized. We can measure electrical activity in the brain's cortex with an electroencephalogram (EEG). Suppose we measure cortical electrical activity with an EEG while an animal is under anesthesia. In that case, we will see brain waves that are characteristically low frequency (also called slow) and high amplitude. They demonstrate the animal is profoundly unconscious under the effects of the anesthetic agent. We can then insert an electrode into a nucleus within the reticular formation while the animal is in this condition. If we begin to stimulate the nucleus electrically, we will see a dramatic change in the EEG recording. As soon as the current goes into the electrode, the EEG will show fast-frequency, low-amplitude excitation in the cortex. This pattern is usually seen only when an animal is wide awake and fully conscious. When we turn the electrode off, the brain immediately returns to the typical slow-frequency, high-amplitude pattern associated with the anesthetized, unconscious state.

This demonstration tells us that general anesthetic agents make surgical procedures painless *not* because our brains are *shut down* but because they are *shut out*. Under anesthesia, our brain cells still work perfectly fine. The systems for feeling pain are all still in place and working perfectly. The pain receptors in our body are still firing. The nerves are all sending barrages of pain signals up to our brain. It is simply that the cortex is completely cut off from all that incoming traffic from the body. It is not dialed into the station.

Here, finally, is the gatekeeper of consciousness. It has become clear the reticular formation is responsible for controlling what signals are permitted to access the conscious centers of the brain. For this reason, the formation in the brain stem that Ramón y Cajal had discovered was renamed the ascending reticular activating system because it is the primary determinant of whether signals are sent up (hence "ascending") to the cortex for processing or not.

Scientists also discovered that if the ascending reticular activating system itself was destroyed in an animal, the creature lost all chance of regaining consciousness forever. More important, the ascending reticular activating system also explained how our subconscious processes are separated from our conscious ones. And it elucidated why Adrian Owen was able to find patients who could awaken because their ascending reticular activating system was intact.

Cortical Blindness: Seeing the Subconscious

Before we go any further, I want to make a second observation about the central nervous system; namely, *the brain rarely works the way we think it does or think it should.* We will revisit this notion many times as we explore the CNS together. Let me demonstrate this notion in action with a clinical example that uncovers the subtle interactions between conscious and subconscious processes in the brain. For this, we need to explore a surprising type of blindness.

Our vision works by transferring light to the retina of the eye and taking the signals from there through fiber tracts that carry it to the very back of the brain. They go to a specific area where visual perception occurs, called the calcarine fissure. This area can be affected by a stroke or hypoxic injury (drowning, smoke inhalation). When this area becomes too deprived of blood flow, the neurons devoted to visual perception die, and the patient becomes blind. This kind of blindness is called cortical blindness.

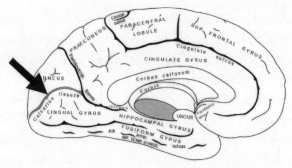

The calcarine sulcus (black arrow) is seen along the medial surface of the left side of the brain.

If we were to examine a patient with cortical blindness together, one of the first clues that would make us suspicious something unusual was afoot would be that the patient's pupils would still work. They would constrict whenever we shined a light on them. Even though the patient could not see the light. But we would see it gets "curiouser and curiouser," as Alice in Wonderland put it.

If we held the light off to the side, we would also see that the patient's eyes would turn toward the light—even though the patient is blind and cannot see it. Now, if I were to toss a baseball at our patient suffering from cortical blindness—I should point out that this is not something I typically do to examine blind patients—he or she would be able to catch it! Even though they were blind and would never see the ball. So, we could dig deeper. We would notice that if we allowed our cortically blind patient to walk unassisted around the room full of furniture, somehow they would mysteriously avoid bumping into any of it. We could also place our cortically blind patient in an fMRI scanner. Even though these patients are blind, we can flash a series of photographs in front of their eyes. We would see that the areas of the brain that respond to facial recognition and the emotions expressed by the faces in the photos (see figure 3) would all "light up" with activity. I know: there are times when facts about the brain become stranger than fiction. But how can we explain these paradoxical findings?

These peculiar results occur because the physiology of our vision and our optical nerves is intact. The incoming visual information from our retinas is still being sent to *subconscious* centers of the brain. Patients who are cortically blind lose their *conscious vision*. All the relatively high-level *subconscious* visual abilities (for example, catching balls, recognizing faces) are still intact. Those signals are routinely processed *before* the individual normally becomes consciously aware of them. Cortical blindness cleaves subconscious from conscious functions. This brings us to another observation about brain function.

Information Overload and Multitasking

We cannot change the underlying functional limitations of how our brains evolved. It is the organ that has been given to us by millions of years of evolution. We must accept that there is little we can do about its inherent restrictions of cognitive bandwidth we mentioned earlier. And that is becoming a new problem for us in the digital age. Never has so much accumulated data been available to so many brains at such dizzying speeds. More than 250,000 tweets, two million internet queries, seventy-two hours of video, and 100 million emails—that is 350 gigs of data—are downloaded, sent, exchanged, and posted *every minute*! And most of us believe we can handle this with multitasking.

You can open all the emails you want while you chat on the phone, text your friends, and scan the newspaper headlines—all while knitting a quilt—and as much as you may think you are multitasking, you are wrong. The Massachusetts Institute of Technology neuroscientist Earl Miller, an expert on multitasking, summed it up: "Switching from task to task, you think that you're actually paying attention to everything around you at the same time. But you're not. You're really toggling between tasks at amazing speeds. Apparently, we were never multitasking. It's a myth!" Now, there is always a small group of people who will say, "That may be true for most people, but I am an accomplished multitasker. I am way more productive when I can dart back and forth between projects in front of me." Researchers at Stanford demonstrated that so-called self-described heavy multitaskers were *worse* at multitasking than people who did not routinely do it. Multitaskers were less efficient at switching from task to task and did a poorer job recollecting information.

A public firm in London commissioned another study of multitasking. Researchers tested the company employees under quiet conditions with items derived from standardized IQ tests. They then presented the same tests to the staff while they were answering telephone calls and email messages. The results showed that the subjects' IQ scores dropped by an average of more than ten points when multitasking. So, you get a little stupider when you multitask too.

The Ants of Consciousness

Neuroscience can often lead one into a kind of neuro-mysticism. The discussion of consciousness directs us to ask one of life's essential questions: Are any of us significant? I am not wondering about what each of us may or may not achieve in our lives. But to consider if individual consciousness might have meaning beyond the confines of the self. We recognize that the cumulative experiences of an individual life are assembled, sustained, and transmitted by our being, as we discussed earlier. But could consciousness mean anything beyond, or outside, that individual identity?

We see the casting of an ant colony. It is the handiwork of millions of ants. It is a marvel of interconnected ventilation shafts, climate-controlled nurseries, factories for the raising and harvesting of nutrient

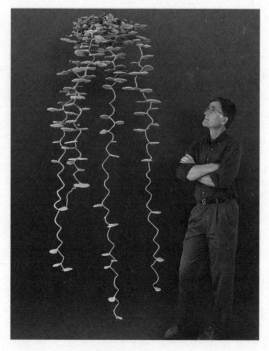

A plaster cast of an ant colony of Florida harvester ants, *Pogonomyrmex badius.*

fungi to feed millions, and superhighways to permit the rapid transport of inhabitants and goods throughout the community. When we gaze at it, we can imagine it to be the manifestation of a single, great, and creative intelligence, say, an architect with the capacity to design a community and infrastructure that can address the needs of millions upon millions of independent insect beings. The sophisticated, complex structure *appears* to be the product of insightful awareness, yet not one single ant involved in its construction had any inkling of the all-embracing significance of the structures in its design. No blueprints. No single architect on site.

Try to imagine the ant's dilemma as it contemplates the world of its hive. If it could ponder, the ant would gaze in wonder at the beautifully constructed arches and chambers of the colony it traverses each day. The ant would have a vague, unsettling feeling that it must be living in a world with a larger significance than it can grasp. This is like our plight as we look out at the universe. We look at it with its big bang and its constant and puzzling expansion—now nearly ninety-three billion light-years across—and experience, perhaps, a sense of wonder like what the ant would experience. He wonders, what is the connection (if any) between my being and this immense construct around me? We look at the universe and ask, who is the architect? Am I a random speck or a directed soul within it?

The Little Blue Dot and Our Big Brain

I was recently watching a television show about the Voyager mission, which began in the summer of 1977. The spacecraft left Earth nearly four decades ago. In 2012, it left our solar system and became the first manu-factured object ever sent out into interstellar space. It has traveled farther than anything in history. Its mission has been extended three times. It took advantage of a singular alignment of the planets in our solar system that happens only once every 175 years. This created a unique trajectory where the space probe could be sling-shot from one planet to the next until it has now traveled more than four billion miles. From that position, it took a momentous photograph (see figure 4) of Earth in its rearview mirror, so to speak, as it left our solar system for good.

Voyager 1 revolutionized our understanding of our own solar system and where life might be found on the planets and moons within it. But it also gave us a context to see what life means on this planet. I looked at the picture and saw what a tiny bluish point our planet represents. The narrator of the program read a quotation from Carl Sagan, from a lecture he gave at Cornell University in 1994, as he showed *Voyager 1*'s historic picture to the audience:

> Consider again that dot. . . . That's us. On it everyone you love, everyone you know, everyone you ever heard of, every human being who ever was, lived out their lives. The aggregate of our joy and suffering, thousands of confident religions, ideologies, and economic doctrines, every hunter and forager, every hero and coward, every creator and destroyer of civilization, every king and peasant . . . Think of the endless cruelties visited by the inhabitants of one corner of the dot on scarcely distinguishable inhabitants of some other corner of the dot.

The late Dr. Sagan was correct. The destruction and atrocities that have happened during human history took place because they were products of a human brain that has evolved to be steadfastly and pathologically fixated upon itself. Until recently, whenever the human mind looked in the rearview mirror, it could see only its own reflection. But now, with *Voyager 1*'s photo from beyond our solar system, the brain must be more aware of the pale blue dot and its meaning. In a collective cosmic gasp, we have a new map where the brain sees its place in terms of cosmic scale.

2

On the Nature of Affection and Trust

The Dinosaur Within

Suncor Energy runs one of the largest tar sands mining operations in the world for the extraction of petroleum. Its biggest mine, called the Millennium Mine, lies near Fort McMurray in Alberta, Canada. In 2011, an engineer was operating a backhoe hundreds of feet underground in the mine when he spotted something unusual. At first, he thought he might have come across a strange rock formation that might have been trapped in the oil-impregnated sandstone. But as he scraped more of the soft stone away, he saw gigantic scales. It looked like the outside of a dragon. The company called in a team of paleontologists to evaluate the find. It proved to be one of the most significant dinosaur discoveries of all time. What had been uncovered were the beautifully mummified remains of a three-thousand-pound dinosaur. The skin, muscle, bones— even the stomach contents—were perfectly intact because of the preservative effect of the petroleum in the sandstone.

It took thousands of hours to fully unearth the remains of a gigantic horned, armor-plated plant-eating dinosaur. It had to be encased in a fifteen-thousand-pound block of plastic to protect the unique specimen while it was extracted and eventually transported to the Royal Tyrrell Museum. The fossil was more than 100 million years old. It represents a previously unknown species that was subsequently named the *Suncor nodosaurus*. But if you look at these remarkable remains, they are so well preserved it seems as if you have stumbled upon the creature while it is taking a nap.

Mummified remains of the three-thousand-pound and 110-million-year-old *Suncor nodosaur,* uncovered in Suncor Energy's Millennium Mine in Alberta, Canada.

The Scaffolding Rule

The brain inside this preserved nodosaur is virtually identical to the hindbrain and brain stem found in you and me. In the same way that paleontologists proceed through the geological strata to determine which creatures inhabited the planet during the past, we must go through the brain layer by layer to understand what parts of it were predominant in ancient times. We are living neuroanatomic fossils: within each of us is distilled hundreds of millions of years of brain development that impart to us our appetites and behaviors. I want to pause here for a moment to introduce you to what I call the scaffolding rule.

The scaffolding rule states that the human brain can evolve new structures and acquire new functions only by building them atop the brain's preexisting anatomy and capacities. Another way of stating this is, you cannot build anything new except atop something old. It is like a building where you cannot tear down any of the existing structures, but you can put up scaffolding to build a new facade in front of it or add more floors atop it.

We can sift through the brain, starting with its newest addition—the *neocortex*. The name is derived from the Greek word *neo* for "new" and the Latin word *cortex,* which refers to the "bark of a tree." The neocortex covers the entire cerebral hemispheres like a new layer of bark.

The surface of the cerebral hemispheres is the contribution that primates brought to the brain's structure as they evolved. This neocortical layer is made up of billions of neurons and is less than four million years old. It is where we do our heavy-duty cognitive processing: language, mathematics, and art. This is also where we develop our plans, make judgments about risk, and execute strategies.

We then dive below the neocortex until we reach the paleomammalian *limbic system* (also called the limbic lobe), which is more than twenty-five times older than the neocortex. From the limbic lobe, we descend into the deepest and most ancient layers of the central nervous system: namely, the reptilian brain—the part we share with nodosaurs. So, our brains are composed of these three distinct brain systems. They form the triune brain. It took more than half a billion years to assemble our central nervous system from three different brain designs: reptilian, mammalian, and primate.

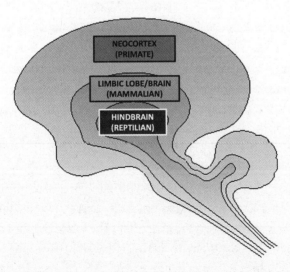

The triune brain: the oldest and most primitive part of the brain is the reptilian one in the hindbrain and brain stem. The limbic lobe is the mammalian portion, and the neocortex is the primate contribution to the triune brain.

The Triune Brain

As you might imagine, reflexive and instinctive behaviors are relegated to the reptilian brain. While it is more than 300 million years old, it remains vital. To understand why, imagine a reptile (say, a crocodile). This will give you a feel for this ancient part of our central nervous system. It runs on instinct. Reptiles lunge at their food. They thrash out violently at the first suggestion of threat or challenge. It's all gut reaction and no thought. As a parent, reptiles reproduce by laying eggs and have little attachment to their offspring. Adult reptiles have been known to devour their young, seeing them as little more than food. Perhaps this should get filed away under "Harsher Reptilian Parenting Practices."

We move upward and outward from our reptilian brain to the limbic system. It is the mammalian contribution to the triune brain and is devoted to emotional processing and output. To understand how mammalian development and the limbic system are yoked together, imagine any mammal you want, but I am partial to using the wolf for illustrative reasons. When a female wolf has offspring, she has a well-protected den in which to birth her young. The pups are born blind and helpless and depend on their mother. There is a prolonged infancy, and offspring often form intense and long-lasting relationships in the pack. Live birth ended up being a hugely successful reproductive strategy but one that required an intense emotional bond to create an enduring commitment, so a mother would be motivated to protect and provide for her young, even at the risk of her own life. These same emotional ties also socially bind members of the pack to each other. While the reptilian brain is focused on instinct, the mammalian brain is concerned with emotion.

The last component of the triune brain is the primate one, responsible for our higher cognitive functions; it is the thinking part of our brain par excellence. One characteristic that sets primates apart from other animals is the large area and thickness of the neocortex. The folding of the cortical mantle upon itself gives the brain its corrugated, walnut-like appearance.

The following observation is derived from the nature of the triune brain itself: the brain stem embodies reactivity, the limbic lobe emotion, and the neocortex cognition. These structures also represent the phylogenetic order in which brain function evolved. The brain stem represents the most ancient structures of the CNS, followed by the limbic lobe, with the neocortex being the brain's newest addition. The most recently acquired functions are also the most vulnerable to disruption. This means that under duress (like a trauma or life-threatening emergency) cognition arising from the neocortex will fail first, then emotional responses generated in the limbic lobe, and, finally, we are left to rely only on the unthinking, reactive instinct of the brain stem. We're back with the reptiles.

The emotional intensity of mammalian parenting is largely driven by a neurotransmitter called oxytocin. It is one of the most important, powerful, and vital chemicals circulating in our brains today because it has been manifesting itself for 100 million years.

Mammalian Power: Oxytocin

This brings us back to the scaffolding rule for a moment. Oxytocin (*not* to be confused with the opiate OxyContin) is secreted by neurons in the center of the brain. Its secretion rises dramatically throughout pregnancy and reaches its highest concentrations right before the mother is about to give birth. Oxytocin initially began as a molecule that interacted with receptors in the muscular tissues of both the uterus to help expel the newborn and the breast to induce the letdown of the breast milk to sustain the infant.

However, as mammals (and later primates) evolved, additional layers of functionality augmented oxytocin's role. In addition to its original peripartum role, oxytocin engendered and sustained the profound emotional connections and attachments that mammalian-style parenting and society required. As mammals evolved, areas of their brains began to develop receptors for oxytocin so the neurotransmitter could affect these regions to orchestrate complex motivational and behavioral reactions. In primates, these responses would shape feelings we would now label affection and trust.

Going back to my wolf for a moment, the nursing alpha female (usually the highest-ranking reproducing female in the pack) must stay in the den to protect her litter of pups and feed them. Meanwhile, she must *trust* (think oxytocin) that the pack will go out and hunt to provide her with the nourishment she needs to feed herself and her young. Alternatively, the alpha female may go out and join the hunt herself, in which case she must delegate the babysitting of the litter to a closely allied, reliable female in the pack (often a sister or daughter of the alpha female). In addition, the alpha female must not only display affection toward her pups but recognize, cultivate, and maintain the complex social bonds that knit the pack around her and her offspring. All these complex relationships are orchestrated and modulated via oxytocin.

Ultimately, oxytocin induces an abiding sense of *connectivity* that frames not only the she-wolf's role as the ultimate guardian of the next generation of wolves but her kinship with the other pack members who share responsibility to safeguard the group's mutual investment in her offspring. And, finally, when all else fails and these pups find themselves directly threatened, little can compare with the intensity of the mother wolf's protective instinct. So powerful is oxytocin's hold that it can lend the alpha female the strength and determination to drive off more than one attacker, against all odds. Again, this is the power of oxytocin. It is the key to the mammalian success story.

Since human offspring have the longest period of parental dependency of any species, it is not surprising that oxytocin's ability to induce behavioral changes reaches new heights in the human brain. Early in the course of her pregnancy, a human mother begins to release substantial amounts of oxytocin from the posterior part of the pituitary gland at the base of the brain. In animal studies, early oxytocin release "primes" complex maternal behaviors and attitudes that significantly enhance both maternal and neonatal well-being. For example, rat mothers who express higher levels of oxytocin release during the first trimester of pregnancy make better and more attentive mothers who are more nurturing and more successful at raising their offspring. This reinforces, in turn, natural selection pressure for the expression of those traits. There is a similar trend in human mothers in whom higher levels of oxytocin are associated with an increased incidence of "mothering behaviors," such as cuddling, picking up the infant, and even checking on the baby more frequently while it sleeps.

Once again, Mother Nature uses evolutionary selection to extend the scaffolding for oxytocin's role. Research in *nonpregnant women* (and, more recently, *men*) has shown it plays an essential role in relationships besides those with our offspring. The mother-infant relationship is the foundational template for the close and intimate relationships we create throughout our lives. To a large extent, oxytocin is the molecule upon which emotionally meaningful bonds and relationships depend. This mammalian "birthing" molecule was "hijacked" in humans so it could also be applied elsewhere. As high as the concentrations of oxytocin are in women during labor and birth, they are superseded by one thing: love.

Romance produces the highest levels of oxytocin release ever recorded. Furthermore, the higher the oxytocin levels measured in the earliest stages of courtship, the greater the odds for success and longevity in the long-term relationship. It has been said that Cupid's arrows are surely dipped in oxytocin.

Oxytocin secretion is not reserved just for these commanding emotional milestones, like birth or courtship. Other things in our lives allow us to experience a "pulse" of oxytocin secretion, too. Almost any act of affection will cause oxytocin to be secreted from the pituitary gland. So-called warm contact hugs (lasting twenty seconds or longer) from a cohabiting partner produce sustained oxytocin release. Even a gesture such as patting a friend's hand or side of their arm, or a consoling hand placed around the shoulder, or a playful pat on your child's head—all promote oxytocin release. Even petting your dog produces a surge in oxytocin levels!

Oxytocin has more magic: it is also the "stay cool and chill" hormone. Oxytocin secretion minimizes the severity of our fight-or-flight response. The body is flooded with a subjective sense of security, safety, and well-being in response to oxytocin secretion. This lowers circulating levels of stress hormones like adrenaline and cortisol. It also reduces cardiovascular stress by lowering blood pressure, heart rate, and respiratory rate. The release of oxytocin also reduces activity in the *amygdala,* a structure in the limbic lobe responsible for generating a sense of wariness and self-protection.

There is a fascinating quality of reciprocity in how oxytocin works: release in one person induces oxytocin release in the other. So, even in casual conversation, if you pay attention to your colleague with genuine warmth and responsive listening, it will produce higher levels of oxytocin in you, the listener, and the speaker as well. And the more oxytocin, dubbed the bonding hormone, is released in the speaker, the more then gets released in the listener. It can continue to create this upward spiral. The "upward spiral of oxytocin," triggered by mutual positive feedback, creates deepening emotional bonds and a sense of trust. This is how love and affection can feel infinite.

Give Us This Day Our Daily Oxytocin

Why is it essential for us to pay close attention to how and where oxytocin release occurs during our daily lives? Because the more oxytocin we release in a relationship, the better that relationship will become. An illustrative example comes from a very "domestic" experiment carried out at the University of Zurich. Before the study began, married couples were asked to generate a list of discussion topics that would create tension between them and often end up in heated discussions. The couples were then brought into the laboratory to discuss one of these touchier subjects. One half of the pairs received a dose of oxytocin via a nasal spray. Oxytocin's molecular structure requires it to be administered either by intravenous injection or via nasal spray. The other group received an inert saline placebo through an identical nose spray.

Researchers then videotaped and observed what happened as the couples tackled the problematic discussions. Couples that had received oxytocin had much less confrontational exchanges. They also exhibited far less stress during the conversation and were more likely to reach a peaceful resolution during the encounter than the subjects to whom saline was administered. I know. You're wondering, when could they start crop dusting the whole planet with oxytocin? This enhanced ability to carry out respectful and cooperative discussions is not just a laboratory phenomenon. Couples who experience long-lasting, satisfying marriages have more sustained, elevated concentrations of oxytocin in both spouses than married couples who reported less satisfaction in their relationships.

Oxytocin is also priming us to scrutinize meaning in our partner's or colleague's facial responses. As we said earlier, our ability to connect and trust others is rewarded with more oxytocin. I have often wondered how many successful meetings rely on oxytocin. Take the famous summit at Reykjavík, where President Reagan spent days in closed-door meetings and having meals in private with the Soviet premier, Mikhail Gorbachev. Was it the closeness of the bond they developed (that is, the upward spiral of oxytocin levels) that led to the most sustained and productive détente since the end of World War II?

A Man Who Understands Intimacy

Daryl Davis is what I would term an oxytocin hero and one of the most remarkable human beings I have never met (but would like to one day). I heard him interviewed on National Public Radio and saw him on Bill Maher's television show. He is a middle-aged African musician. One night he was playing some rock-and-roll tunes on the piano with his band. After they finished their last set, a white man came up and shook Mr. Davis's hand.

Meaning to pay Mr. Davis a compliment, the stranger commented, "You know this is the first time I ever heard a black man play piano like Jerry Lee Lewis." Mr. Davis tried to explain that Jerry Lee Lewis had learned his technique from Black artists who played the blues, jazz, and honky-tonk. The white man told him he was wrong: Jerry Lee Lewis, he insisted, had invented that style of playing the piano.

It occurred to Mr. Davis that there were not many opportunities in his life when he could sit down with a white man, share a meal, and have a real, genuine conversation about music. He decided he would start a conversation with this man about black artists like Fats Domino and Jelly Roll Morton. They talked about music for quite a while and gradually felt more at ease with each other. As they spoke, the white man volunteered that he was a member of the Ku Klux Klan and it was a weird sensation for him to be having such an engaging conversation with a black man! The man went into his bedroom and emerged with his Ku Klux Klan robes and handed them to Mr. Davis, explaining that he no longer had any use for them. This incident became the beginning of a personal crusade on Mr. Davis's part.

Daryl Davis made up his mind that he was going to travel and get to know other Klansmen. He would engage them, befriend them, and see if they could change their attitudes about Black people. If they got to know him as an individual—as a single black man—would they, at some point, be willing to forsake the Klan? The results have proven to be remarkable. To date, more than two hundred Klansmen have surrendered their Klan gowns and hoods to him. The reason I found Mr. Davis's campaign stirring is that it recognized the enormous value of *intimacy*. It harnessed the notion that we create the best opportunities for more meaningful communication when we fashion circumstances that induce and sustain mutual oxytocin secretion.

Daryl Davis invested his time in these Klansmen. He had to get close to them. He had to listen to their tirades about ethnicity, to their biases and prejudices, and, occasionally, to their outright denigration of the trials and tribulations his race has had to endure since the dawn of slavery. Still, he listened. He listened wholeheartedly, profoundly, and respectfully. He listened because he had an unshakable faith in the humanity of each Klansman he met. He had to look deep into their faces, hear their

Daryl Davis is an Black virtuoso blues pianist, author, actor, and lecturer. He is on a mission. Throughout the last thirty years he has befriended members of the KKK and persuaded them to leave their racist beliefs and organizations. Daryl does it through dialogue and music. The U.S. embassy in Tel Aviv invited him to speak as part of a multicultural exchange program.

voices, read their emotions, and see their perspective, biased as it might have been. But he knew that if he could win their friendship, it would become virtually impossible for them to hate blacks because they were already friends with one. This is one of oxytocin's great benefits: *the more oxytocin we engender in someone, the harder it will be for that individual to see us as an enemy.* In general, it is hard to hate someone you get to know intimately. It is Mr. Davis's sustained ability to draw out oxytocin, even in those who might initially look down on him because of his color, that helped those Klansmen to rise above their biases.

When we grow close to other human beings, we see there are common values and grounds on which to build understanding and even trust. I often thought if you want to sit down and negotiate a peace treaty— say, between the United States and North Korea—send in a man like Daryl Davis. Let him hang out and talk to Kim Jong Un. According to the retired NBA star Dennis Rodman, who has a close relationship with the North Korean leader, Kim has children whom he loves very much. Let our diplomatic negotiators invest the time to get to know him and his family the way Rodman did. Let Kim also have the opportunity to know his American counterpart's family. Bring their children together. Let them discuss common interests in art, sports, and music. Let them go to a basketball or baseball game together. Share a meal in each other's home. Not a diplomatic banquet. A home-cooked meal with each other's family eating around the dinner table. Let them grow close and trusting enough until they can share their moments of success and triumph. But, probably more important, failure and shame. Let them live side by side. Then let them negotiate. Oxytocin will show them a way to find peace.

The Brain Needs Love

Nicolae Ceaușescu was a ruthless dictator who ruled Romania from 1965 to 1989. He was considered one of the most brutally repressive leaders in the Soviet Communist bloc. After Ceaușescu's downfall, it came to the world's attention that his regime had created a vast infrastructure of government-run orphanages that contained and processed hundreds of thousands of children over decades. Ceaușescu's national policy was that it was better to raise children in the government's system of orphanages than leave it to their impoverished families. Unfortunately, this approach of raising children in orphanages on an industrial scale demonstrated that addressing the need to be loved is the most crucial ingredient in raising children. It provided tragic evidence that depriving children of affection will inevitably sabotage their brain development.

The Romanian orphanage system has rightly been termed "the slaughterhouse of the soul." After the fall of the Ceaușescu regime, investigators discovered grim confirmation that children raised in the orphanages spent most of their lives confined to a crib and were deprived of virtually any human contact. Members of the orphanage staff had only brief physical contact with the children to change their diapers or feed them. The Romanian orphanages serve as a grim experiment in what happens to babies when they are denied human warmth, nurturing, and affection early in their development. The consequences were more far-reaching than any researcher could have imagined.

The first set of findings about children from the orphanage system was that one in ten suffered from such severe cognitive and behavioral impairment that they would have to spend the rest of their lives confined to a residential institution.

The next surprising finding in the evaluation of these children came in 1999, when Dr. Charles Nelson, a professor of pediatrics at Boston Children's Hospital, began visiting the Romanian orphanages in person. He found a high percentage of the children raised in the orphanages exhibited profound autism. Many of the children showed evidence of irreparable language impairment. Some could not speak at all. Other children were unable to recognize faces or respond to caretakers. These findings reinforced a hypothesis that a failure to stimulate and sustain adequate oxytocin release might play a role in certain forms of autism.[*]

Nelson's team returned to Romania several times and began collecting more data. A sizable number of the EEGs of the orphaned children exhibited aberrant brain wave activity. MRIs and PET scans were performed on the most impaired children. The neuroimaging findings were alarming: almost all severely affected children showed dramatic reductions in

---

[*] There are many outdated theories that have been debunked by hard science but continue to circulate—namely, that autism spectrum disorder (ASD) is related to "bad" parenting. This is not the case. In the case of ASD, the hypothesis is that the oxytocin receptors are dysfunctional and/or inadequate in number.

gray and white matter. The volume of the children's brains—the numbers of brain cells and the width of the interconnecting fiber tracts—had been tragically and irreparably reduced. The children's brains were profoundly underdeveloped from the lack of affection, warmth, and parenting. A dearth of oxytocin had damaged their brains. The brain scans revealed more bad news: children who were fortunate enough to be adopted into supportive homes before the age of two appeared to be able to go on to develop normally and make good recoveries. But children over the age of four were doomed. Their brains could no longer catch up or repair the maldevelopment.

It is hard to see these findings. However, history has always brought us monsters who conducted their grim social experiments by placing

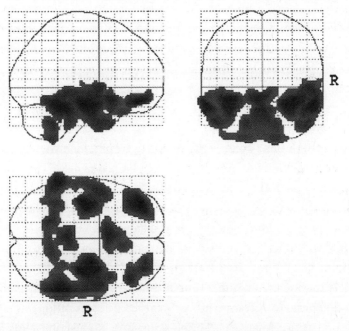

These are composite images showing dramatically decreased brain activity in a group of Romanian orphans compared with a control group. The dark areas are all significantly diminished activity as seen in all three views of the brains of Romanian orphans. Years of neglect along with a lack of affection have left the orphan brains grossly underdeveloped in all major cortical areas, especially the cognitive area of the frontal lobes and the temporal lobes where emotional processing is carried out.

the state's needs over the well-being of its offspring. Like Hitler and Ceauşescu. One of the most loathsome of such leaders was Frederick II of Hohenstaufen, who would later become the emperor of the Holy Roman Empire in the mid-thirteenth century. He was infamous for his sadistic streak. He became obsessed with many things, but among them was discovering "humanity's original language"; that is, what language did Adam and Eve speak? To answer this question, he had newborn infants rounded up and placed in the hands of nurses whom he had specially trained. They could handle the child only when necessary, such as to bathe, clothe, or feed them. But under no conditions were these nurses to hold the babies, caress them, or communicate with them in any way or fashion. He believed that if the children were maintained under these harshly constrained conditions, they would eventually be compelled to speak the "original" language. The emperor ran his experiments several times. He never succeeded in getting the answer he sought because every single child died! They died from a lack of affection. They died because oxytocin was missing in their short lives.

Love Is the Template

The maternal-infant bond is the template from which we trace all our social and emotional connections throughout our lives. The necessary warmth and nurturance need not be derived from the mother per se. It can be a father, or a grandparent, or a member of the extended family. Without parental love (or a meaningful substitute) in infancy, however, the child will suffer a lifetime of brain maldevelopment and dysfunction.

Brain imaging reveals a stark truth: the first three years of life represent a uniquely vulnerable time in brain development. This finding has led experts to recommend early targeted interventions in susceptible children to enhance parenting skills, nutritional counseling, interventional and preventive health care, child protective intermediations, and enhanced learning and educational readiness training. So, it would be fair to ask, what are the results of such early preventive childhood programming?

The RAND Corporation did an exhaustive review of the effectiveness of such targeted childhood interventions: all of them enhanced the emotional and cognitive capacities of the children receiving the support services. There were improved educational outcomes and school performance. There were also tangible results in improved economic self-sufficiency in the parents of these children. And these children themselves went on later in their own lives to show increased self-sufficiency, lower high school dropout rates, more participation in the labor force, improved incomes, and less use of welfare services compared with children who had not received these early childhood support services. Children receiving support services were less likely to engage in criminal activity and end up in prison. Finally, health and well-being measures were increased later in their adult lives, and they needed fewer health-care services. The recipients of these programs were all followed well into adulthood. The benefits of these early childhood interventions even carried over to the next generation. The cost savings to society and government were substantial and irrefutable: for every dollar spent on targeted childhood interventions between birth and the end of preschool, there was a return of $3 to $5 saved. Affection and support are more than their own rewards. Their beneficial effects ripple throughout society and across generations.

In 1935, the U.S. government passed Aid to Families with Dependent Children (AFDC) as part of the Social Security Act. AFDC was meant to provide financial assistance to unemployed families who had children. In 1988, the AFDC was expanded to include children whose parents had difficulty supporting them because of disability, lack of employment, or being single parents. Most voters assumed the vast majority of recipients of the AFDC program would have been Black people. That was wrong. Most beneficiaries were, in fact, white.

In 1996, under sustained conservative political pressure, largely led by the then Speaker of the House, Newt Gingrich (who had publicly criticized entitlement programs like the ACDF as part of his "Contract with America"), Congress repealed the AFDC program and replaced it with the Temporary Assistance for Needy Families (TANF) program. Not only did TANF reduce the level of support for these children across the board, but it also made the assistance temporary: it would last six months and then be withdrawn. The unintended outcome was that under the TANF program more children were placed into foster homes because their parents ran out of financial assistance. Short-term temporary funding for low-income families turned into a mill for funneling children into foster care. As one report summarized the effects of TANF, "In 21 states, this ratio [of children in foster care versus children at home] is greater than .5, meaning that for every two children receiving TANF while living at home, one or more children are living in foster care. In seven states, the ratio is 1:1 or more, and in two states (Wyoming and Indiana) there are roughly two or more children in foster care for every one receiving TANF while living at home." In many states, the number of poor children placed in foster care soon exceeded the number of children who still lived at home. TANF has ended up being a shortsighted wrecking ball—pure and simple—that has destroyed families and separated children from their siblings and their parents. It also was passed in the face of years' worth of evidence of the wisdom and effectiveness of the AFDC. Government still trumps childhood.

The Enduring Effects of Childhood Trauma

Scientists are now discovering that the chronic effects of neglect and childhood trauma (CT) during the first decade of life are far more pervasive than previously thought and can manifest themselves well into adulthood, last an entire lifetime, or even extend beyond the life span of the abused individual. The traumas involved included maltreatment by adults and peer bullying. Although CT has long been associated with significant adverse mental health outcomes, there now appear to be epigenetic mechanisms that underlie the onset of delayed psychiatric illness. These include modifying receptor sites for the binding of gluco-corticoids; the latter have been implicated in the harmful effects of stress and the increased incidence of depression. Evaluating a large group of children referred to Child Protective Services for documented maltreatment demonstrated they exhibit higher levels of inflammatory markers in the bloodstream than age-matched controls. Additional epigenetic mechanisms that affect the health and maintenance of healthy neurons were also seen. These latter mechanisms include elevated inflammatory biomarkers in the central nervous system itself, including elevated numbers of T-lymphocytes and pro-inflammatory cytokines (for example, interleukin-6). The latter are known to play a role in interfering with neuronal connectivity, accelerating dendritic pruning, and diminishing neuronal plasticity. There was also significant mortality and morbidity associated with suffering abuse, neglect, or trauma in childhood; as such, individuals suffered from a higher incidence of cancer, heart disease, and diabetes.

The epigenome is the pathway through which nature and nurture meet. While so-called epigenetic changes do not represent actual mutations of genes in the DNA, they represent substantive changes in the ability of the DNA to be expressed that have been the result of environmental influences experienced by the individual. And epigenetic changes can be passed through several generations beyond the affected individual. So, we see two significant developments here. The first is that CT puts the child at higher risk of developing both physical and mental health impairments. And the second is that the trauma we suffer as children becomes an intergenerational issue. The wounds we acquire as individuals during childhood may be passed along to our children and our grandchildren. Recent research has also elucidated that trauma can affect the fetus in utero. If a mother suffers intimate partner violence during pregnancy, this exposure can induce epigenetic changes in the fetus without the mother ever experiencing the epigenetic changes herself. The significance of this finding is that it impresses upon us the unique vulnerability of the developing child to violence and trauma extending all the way back to the womb. Whether an individual's epigenetic changes can be reversed or not is something that is beginning to receive scientists' attention but has still not been determined. In the meantime, these early findings about the epigenetic sequelae and intergenerational consequences of CT are ominous enough that they suggest we should take every reasonable precautions to ensure that sufficient protective and interventional safeguards are in place in our our schools and our communities to mitigate or prevent CT wherever possible. If we do then we will be saving not just the child that stands before us now but his or her yet unborn descendants to come inot the world in the future.

When we look at society's ills, like greed, global conflict, class strife, inequality, prejudice, and indifference, most are attributable to our inability to generalize the template of love and apply it to the world at large. As

The plight of refugee children (who currently number more than forty-three million) poses serious questions about what developmental, cognitive, and emotional scars and deficits we may be inflicting on their developing brains.

I write this, the United Nations Refugee Agency has estimated that there are about one hundred and ten million refugees globally; forty-three million of these are children. They are the innocent victims of genocidal wars, starvation, political secularism, and religious persecution. Many themselves have been targeted for destruction in their own homes, schools, and hospitals as a way of terrorizing the local population. As the refugee crisis worsens, particularly with the sectarian wars throughout the Middle East, Africa, and Asia, the question is *not,* how much will it cost to take care of these women and children? No, rather it should be, can we afford not to?

The Oxytocin Web

Oxytocin is the neurochemical equivalent of Spider-Man's web. It can stick to anyone. Shore up a relationship. Or draw its subjects closer. A study was recently done at a family wedding. In this case, the researchers had obtained permission to draw repeated blood samples from the attendees. I hope the scientific team brought a big wedding present.

I try to imagine what the wedding might have looked like with a band of lab technicians scurrying around, armed with syringes in one hand and carrying a tray of hors d'oeuvres in the other. The servers might ask, "Would you care for a canapé and would it bother you if we took another blood sample before the next toast to the bride and groom?" As the wedding proceeds, lab techs are drawing blood samples like maniacs, labeling them, putting them on ice, and going back for more. Given this ludicrous scene, let me pose a question: Among all the people attending the wedding, in whose bloodstream do you think researchers recorded the highest levels of oxytocin?

If you said the bride and groom, you were right. Okay, so where do you think the second-highest levels were measured? The parents of the bride and groom. Right again. And the rest of the crowd were all connected by oxytocin levels that were predictably based on how closely tied the test subjects were to the couple getting married. Lowest levels? I would guess the bartender and caterers had the lowest. Just a gig.

3

On the Nature of Romance

Tonight I can write the saddest lines.
I loved her, and sometimes she loved me too.
—PABLO NERUDA1

Although the poets may insist that love is a matter of the heart, it is not. It is a matter of the brain. If there was ever a place in human affairs where brain imaging and neuroscience have fundamentally changed how we understand an emotion, it is romantic love.

Despite its primacy, romantic love is hard to define. Dr. Sarah Pinto is a historian at Deakin University in Australia. She has studied societal notions of romantic love across different cultures and at different times in history. She points out that while "some people would say romantic love has to involve a sexual or erotic component," it has proven to be far more mercurial and dynamic. "It's not something that is always the same in different times and places. It's a cultural phenomenon that changes."

Our current views on romantic love, for example, emerged from the Middle Ages. At the time, love was envisioned primarily as part of a larger moral and religious struggle to refine one's soul. It called upon the individual to endeavor to rise above erotic desire and abide by a loftier code of chivalry. In the nineteenth century, courtship and betrothal shifted away from arranged marriages built around political alliances to less constrained relationships increasingly driven by mutual attraction and passion. The concepts of erotic and romantic love changed with this shift in emphasis.

The Chemistry of Romance

We just reviewed how maternal love, fueled by oxytocin, serves as the template for our future relationships in life. As we grow into adulthood, our quest for love changes from one based on security to one based on sexuality. And the brain enlists a new group of neurotransmitters to carry out this assignment.

We could group these neurochemicals under the general heading of "the feel-good neurotransmitters." The first one is called dopamine. Its role in the brain is so powerful that at times it can seem almost tyrannical. Dopamine crops up in nearly everything the brain does, but it has two dominant functions. The first is to determine what, where, and when we pay attention; that is, what is important to us? What are we motivated to pursue? The second is producing pleasure or, more specifically, a reward. Dopamine is the foundation of the brain's entire pleasure and incentive system. The pursuit of gratification is a powerful force in our lives.

Dopamine is central to understanding the nature of overwhelming desire and even outright addiction because the neurotransmitter can motivate us to pursue an objective to the exclusion of everything else, even if our actions might be self-destructive. To understand why dopamine is so powerful, I want to dissect some of its various effects. The first of dopamine's signature characteristics is that it induces actions that are tinged with anticipation. That sense of "I'm dying to have a cookie" or "I must have her back in my arms again" or "I got to get another hit of heroin"—these are all sentiments dripping with dopamine's influence.

One of the telltale signs that dopamine is at work is that our sense of yearning is a more powerful component of our motivation than the actual "getting." For now, I want to use a straightforward example that does not involve romance to illustrate these features. Let's say we've been having terrible cravings for chocolate cake. And we'll assume chocolate cake has been on our minds now for several days, and it just won't quit. It keeps coming back to nag us.

At the end of our workday, we get off our bus and walk the three blocks home. Our thoughts drift back to a delicious slice of chocolate cake. Hmmm. Suddenly we realize we have absentmindedly walked a block too far, and we now happen to be standing in front of our neighborhood bakery. We spot a beautiful slice of chocolate cake right there in the bakery window. We scratch our collective heads and wonder, "How the heck did we end up in front of the bakery?" Short answer: dopamine.

This part of the vignette emphasizes some of the qualities dopamine lends to our motivation. The first is dopamine creates recurrent yearning. Each time it orbits back and commands our attention, there is a greater sense of urgency. That is characteristic of dopamine: once it begins to drive our behavior—especially our *seeking* behavior—it tends to keep urging us ever harder toward gratification. This is what magically guided our footsteps to the bakery.

We tell ourselves, "Well, we're already here, right?" So, we walk in and buy a slice. The baker behind the counter asks us, "Would you like me to wrap this up to go?" "No," we answer, "we'll just eat it here." Of course, we will! Dopamine is going to make sure we do. We sit down and take our first bite. It is pure heaven! Hmmm! Then we take a second one. Hmm. It's still really good but not *quite* as delicious as the first one. And by the time we get to the tenth bite, we might even be ready to quit and push the plate aside.

An essential function of dopamine is to create *anticipation*. It is directly related to the excitement we feel just as we get ready to eat our first bite of cake. We will learn to recognize when we feel that tension, we are under the effects of dopamine. The quality of what we might term successive disappointment—each successive bite is less tasty than the preceding one—is another of dopamine's properties. The first hit, the first taste, the first kiss, is always the most intense. And then, with each repetition, the intensity fades. This "tolerance" is an integral part of dopamine's role in addiction. Because the reward derived from the brain's pleasure centers is always at its most powerful the first time around. The second is always less. Unless we can crank up the second reward—jack up its intensity—it is bound to disappoint. It will always be less rewarding than the first "hit."

The next quality of dopamine is a gradual built-in *tolerance* inherent in the dopaminergic system. This is because, even moment to moment, the dopamine receptors on brain cells dampen their responsiveness to each discrete amount of dopamine that is released in the brain. So, whether it is a yearning for a dose of heroin or seeking to overcome unrequited love, dopamine always makes us want more. The addictive behaviors and disorders fueled by dopamine drive us to regain the intensity of that first experience. As the brain's primary intrinsic reward system, dopamine's characteristic properties include motivating us through increased yearning, an escalating sense of anticipation, and building up a tolerance to its effects that makes us compensate by increasing the intensity of the reward.

Another "feel good" neurotransmitter involved in love is named endorphin (or plural endorphins), which is the brain's own "morphine," or opioid. It is because our brain possesses its own endorphin receptor system that exogenous opioids (like morphine, heroin, or oxycodone) not only have their painkilling (analgesic) effects but also produce a sense of euphoria.* You know that expression, "Oh, he's in love. He's three feet off the ground"? It is romantic love's ability to induce high levels of endorphin release in the brain that gives us a powerful sense of elation.

There is one final "feel good" neurotransmitter to touch upon named serotonin. I am oversimplifying things a little, but serotonin governs emotional responses, including happiness and well-being, somewhat analogous to the effects of oxytocin. There are four powerful, primary neurotransmitters associated with romantic love—oxytocin, dopamine, endorphin, and serotonin—and that cocktail of neurotransmitters makes love the most powerful emotion we experience.

The Pursuit of Love

When it comes to romance, our ancestors were lovers. Natural selection favored individuals who could fall in love because it served as a solid foundation for intimate relationships. Lust and romance drew men and women together into devoted mating pairs that could reproduce more successfully. That bond also provided a more stable parenting structure because parents who love each other would be more likely to

---

* Just to make sure we keep the terminology absolutely clear in these days of "opioid" crises, an *opioid* is any molecule, synthetic or natural, that binds to opioid receptors in the brain, while an *opiate* is technically a substance derived from plants occurring in nature, like the poppy. Medications that are heard about frequently in the press like fentanyl, hydrocodone (sold under the trade name Vicodin), and oxycodone (sold under the trade name Percocet) are all termed synthetic opioids.

share resources and responsibilities. However, as we will see, love's power extends far beyond its application to pair bonding. Once again, the scaffolding rule is at work here: the brain's original application of so many rich neurotransmitters associated with rewarding reproduction was to help ensure more effective mating.

Romantic love is, therefore, hardly a neurological accident. In the long arc of evolutionary time, romance became a stunningly powerful emotion because it ended up being a successful and productive biological strategy. Does love lose its magic when we look at it as a bio-evolutionary force? Hardly. Consider this: most individuals who love one another deeply also claim they would gladly sacrifice themselves for those they love. Love not only serves as a method for enlisting behaviors to support reproduction but also lies at the heart of our notions of heroism and self-sacrifice.

A passage from the Bible says, "*Greater love* hath no man than this, that a man lay down his life for his friends" (John 15:13; emphasis added). When recipients of the Congressional Medal of Honor—the country's highest decoration for valor on the battlefield—are interviewed, there are two things you will hear almost universally. The first is, "I'm not a hero. Any soldier would have done the same thing under the circumstances." The second is, "I did it for the guys in my squad who fought alongside me."*

Love gives human beings the capacity to perform heroic acts of bravery, devotion, and self-sacrifice. The recipients of the Congressional Medal of Honor proved themselves willing to sacrifice their own lives to safeguard those of the members of their units. Although love might have originated as a biological principle, it becomes a moral one when it is extended to include our fellow men and women. It has morphed over time from a force reinforcing reproduction to an engine of devotion and valor.

We know romantic love changes as we experience it. Moreover, brain imaging demonstrates that there are profound differences between how men and women respond to love. Professor Helen Fisher, a leading

---

* To date, the Congressional Medal of Honor has been awarded to only one woman. Her name is Dr. Mary Edwards Walker, an extraordinary surgeon who served courageously with the Union army during the Civil War.

researcher on the topic of romantic love, has broken it down into three stages—one could even call them options.

Understandably, lust is usually the first stage; it springs from the secretion of sexual hormones in the brain. Sexual hormones such as estrogen, progesterone, and testosterone play a significant role in promoting sexual drive—the reproductive imperative. The second stage is attraction; namely, what we would traditionally call romantic love. As we will see, this stage has many of the same obsessive characteristics as addiction because the brain's dopaminergic system also fuels it. The final phase in this progression of romance is attachment. Attachment provides a profound sense of friendship, partnership, and communion. This last stage is imbued with the warmth and afterglow provided primarily by oxytocin release. It brings with it a realignment of values in the relationship. Erich Fromm, the great twentieth-century social psychologist, wrote, "Immature love says: '*I love you because I need you.*' Mature love says: '*I need you because I love you.*'"

This brings us to two observations. The first is that romantic love typically evolves in three stages, lust, romance, and attachment, that correspond to the predominant influence of sexual hormones, dopamine, and oxytocin, respectively. Naturally, no stage depends solely on one single neurotransmitter, nor is one stage necessarily exclusive of another. Think of neurotransmitters as dominant themes within a given step. Understanding the effects of various neurotransmitters is becoming an increasingly tricky assignment because more than one hundred neurotransmitters have been identified in the human brain. And every year brings us the discovery of one or two more. When we begin to use brain imaging studies, we see that each stage of love corresponds to the release of a subset of neurochemicals and specific areas in the brain being activated by their secretion.

The second observation is that it is not enough for us to know what neurotransmitters are involved in a particular activity or behavior; we also need to assess what regions of the brain are affected by the presence of these neurotransmitters. For example, we know that romantic love is associated with high levels of dopamine release. Additionally, brain imaging shows us that the hippocampus, an area of the brain where memories are stored, is selectively activated by circulating dopamine.

This means that when romantically charged memories are created in the presence of dopamine, they are tucked away with a high level of romantic appeal. This is why individuals often look to past recollections for ways to rekindle passion. So, one might be tempted to return to the same vacation spot, wear the same black dress, or eat out at the same restaurant to evoke the earlier romance associated with the original, corresponding memory. We will see that this "dopamine tag" associated with laying down romantic memories in the hippocampus can become problematic when it takes on an obsessive quality.

Lovesick

The early phases of romantic love are often tinged with a sense of longing. Sometimes, it can approach addiction, especially because of the dopaminergic (meaning "related to dopamine") underpinnings of the initial phases of romance. We spend many waking hours yearning for the object of our desires when we first fall in love. Helen Fisher sums up the obsessive nature of falling in love as "someone is camping in your head." If we go for too long a period without contact with our beloved, the yearning can physically take hold of us. We can experience malaise, a lack of appetite, and a loss of engagement. We have appropriately coined the term "lovesick" to describe this state.

We become "love junkies" suffering from withdrawal. Two potent dopaminergic areas of the brain are the *ventral tegmental area* and the *nucleus accumbens.* Both are involved in the brain's reward system and are active when we fall in love. The activity detected by fMRI in the ventral tegmental area and nucleus accumbens when we are in the throes of romance equals—or surpasses—the activation seen with a dose of heroin or cocaine. Both drugs have a powerful propensity for addiction. And, like any addictive substance that harnesses these dopaminergic pathways, love can make us feel delirious.

The *National Geographic* writer Lauren Slater summed up dopamine's role in romantic attachment:

> In the right proportions, dopamine creates intense energy, exhilaration, focused attention, and motivation to win rewards. It is why, when

you are newly in love, you can stay up all night, watch the sun rise. . . .
Love makes you bold, makes you bright, makes you run real risks,
which you sometimes survive and sometimes you don't.

However, we can also find ourselves plunged into withdrawal when
we are abruptly cut off from our "drug." We see this, for example, when
someone is suddenly rebuffed by someone they love. The rejection can
plunge the individual into an abrupt withdrawal that can be so intense
and excruciating that it can lead an individual to consider homicide or
suicide. Falling in love has many of the same qualities as drug addiction—
namely, the need for increasing "doses," sustained longing, and withdrawal and despondency when the supply is abruptly stopped. Love can
make us feel a "rush" or "high," and those feelings do not necessarily
fade over time. Neuroimaging demonstrates that even after decades of
togetherness spouses in successful, long-lasting marriages exhibit the
same elevated levels of dopamine as do newlyweds when they look at
photographs of their respective spouses. The thrill brought on by dopamine seems to be the secret to enduring marriages and partnerships: it
still feels great (if not better) after decades!

Falling out of Love

But what happens to our brains when we lose love? To answer this question, Fisher's research team enrolled forty-nine subjects for brain imaging from three separate groups. The first comprised individuals who had
recently fallen in love, and their love was reciprocated. The second group
was composed of subjects who had also recently fallen in love, but they
had just been "dumped" by the person they loved. And the third group
included subjects who were in love and had remained happily married,
on average, for a period of twenty-one years.

The imaging study demonstrated *no* differences in the levels of brain
activity between those who had just fallen in love in the last three months
and those who had been happily married for decades. As far as the fMRI
scans were concerned, love was love—whether it was two teenagers making out in the backseat of a car or an elderly couple sitting in their paja-
mas at the breakfast table.

A second finding might appear paradoxical at first glance. The brain activity detected in the group that had been rejected was no different from the other two groups that had experienced either short-term or long-term requited love. Dejected or exalted, brains in love produced the same images. At first glance, we might object and say, "Wait a minute, one individual is walking three feet off the ground because his love is reciprocated, while the other is languishing in despair because his love is rejected! How can their brains look the same?" Well, the individual's brain is still experiencing love. This is a reliable and reproducible finding: a brain in love looks the same whether that love has been rejected or reciprocated.

There were, however, some subtle, telling differences that separated the brain scans of those who were rejected versus those who were not. The rejected lovers began showing *even more activity* in the areas associated with romantic love. In other words, once rejected, they loved *harder*. We saw activity in the dopaminergic system was ramped up.

Remember earlier, I had pointed out that dopamine plays a role in determining where we direct our attention. So, what is happening in the brains of these rejected lovers? Dopamine makes them train more of their focus on the person they love than before. Thus, dopamine can make the nature of their unrequited love more compulsive. When one sees the soaring levels of dopamine activity in the brains of rejected lovers, it makes their anguish more poignant. Even their brains look as if they ache for love. This is a finding confined to rejected lovers: increased dopamine output associated with rejection can fuel an obsessive component to behaviors observed in unrequited love.

Another difference in brain activity that one sees in studying the scans of rejected lovers is a dramatic rise in activity in the *prefrontal cortex*. This is an area of the brain where executive planning, risk assessment, and strategizing take place. A neurosurgical friend of mine once called the prefrontal cortex "the James Bond part of the brain." He reminded me of all the scenes where James Bond is planning his escape, calculating how long he has until detonation, or deciding if his Aston Martin can zip past the bad guys. All those kinds of activities occur in the prefrontal cortex. Rejected lovers often begin to calculate what measures might be undertaken to win back the object of their affection.

There's a hilarious take on this kind of "rebuffed mathematical reassessment" in the movie *Dumb and Dumber,* where the character Lloyd (played by Jim Carrey) is in love with Mary (played by Lauren Holly). Mary has absolutely no romantic interest in Lloyd. Dejected, Lloyd decides to confront her:

LLOYD: What do you think the chances are of . . . [us] ending up together?

MARY: Well, Lloyd, that's difficult to say. I mean, we don't really . . .

LLOYD: Hit me with it! Just give it to me straight! I came a long way just to see you, Mary. The least you can do is level with me. What are my chances?

MARY: Not good.

LLOYD: You mean, not good, like one out of a hundred?

MARY: I'd say more like one out of a million. [*Long pause.*]

LLOYD: So, you're telling me there's a chance . . . yeah!

So, rebuffed lovers will take chances—even hopeless long shots—to regain their love. *If individuals are rebuffed in a romantic relationship, their unrequited love can engender desperate planning and risk taking to win back the object of their affection.*

The loss of love from rejection is a different case from when one's partner passes away (we will cover grief in a later chapter). However, we can already make some predictions about a grieving lover. First, do you think their fMRI scans show that they are still in love? Yes, of course. When a person passes, we are still in love with that individual. Will we see an increase in prefrontal activity? No. Because we do not expect them to scheme in any way to bring the departed back. We will see, however, that grief has its own signature pattern of brain activity in neuroimaging studies.

Obsessive Love

The activity we detect in the prefrontal cortex of rejected lovers can have a darker side. As discussed earlier, romantic love can have a focused, obsessive quality, and rejected lovers demonstrate regional activity similar to what is seen in obsessive-compulsive disorder. High dopamine output along with increased prefrontal activity can make a rejected lover more susceptible to succumbing to a hypervigilant, compulsive state of mind—one that can turn a suitor into a stalker.

The case of E.F. illustrates the potential for romantic interest to devolve into more dangerous infatuation. While attending college, E.F. became enamored with a young woman named C.W. His love was, however, unrequited. Ten years after graduation, C.W. was happily married, but she began receiving emails, letters, and cards from E.F. Not a few. Hundreds of them. Next, security cameras detected E.F. snooping around C.W.'s home. When this incursion was brought to the police's attention, a warrant was obtained to search E.F.'s residence.

When the authorities investigated E.F.'s home, they found a downloaded photo from C.W.'s wedding serving as the screen saver on his computer. A review of E.F.'s digital records revealed he had carried out more than forty thousand separate inquiries over the internet about C.W. in a single year! He had also placed hundreds of phone calls to C.W.'s home or cell phone. He had broken into C.W.'s car to steal mementos. And at one point, E.F. had even posed as a soon-to-be-parent at the school that C.W.'s children attended.

In this age of social media, stalkers (like E.F.) have found it easier to insert themselves into the lives of their victims and create more distress and anxiety through their digital presence. Stalkers can uncover personal data and photos, becoming empowered to inflict emotional abuse and psychological harm on victims and their families, at their workplaces, among their friends, and in the communities where they live. Eighty percent of victims have a history of a prior relationship (*not* necessarily a romantic one) with their stalkers, who have proven remarkably patient, methodical, and ruthless. The sad news is that stalkers will victimize one out of six women in the United States.

Some stalkers have carried out the pursuit of their victims over not just years but even decades, despite restraining orders and other legal protections. Even with current statutes, restraining orders still allow stalkers leeway to hound their victims through the use of the internet and social media, making the old-fashioned avenues of restraint based on physical proximity irrelevant. Ninety-eight percent of us will experience unrequited love. Thankfully, stalking occurs in a small minority of relationships and is usually carried out by individuals with a history of mental health issues. Nonetheless, it is an example of the obsessive, dark turn that romantic love can take.

Love at First Sight: Finding Treasure on the Map of Love

Some aspects of love seem to seize us instantaneously. In *Hero and Leander,* Christopher Marlowe wrote, "Who ever lov'd, that lov'd not at first sight?" How and why does it happen? One clue lies in the way our brains create "maps" of what our ideal mates should be. We fashion a template of the person we imagine as our future partner. It is interwoven with innumerable sources from our personal experience. Parts of the template are modeled upon relationships in the first years of our childhood. Other aspects are shaped by cultural norms in advertising or entertainment. In our minds, we create an unwritten, subconscious set of characteristics our perfect mate should exhibit. If we encounter someone who matches many of these qualities, we begin to feel a deep attraction and romantic interest. If they are a very close match, we may fall in love at first sight.

It turns out that falling in love at first sight is not a bad way to select your mate. Love at first sight works. In one long-term study, 75 percent of those who fell in love at first sight went on to develop a long-term relationship with that person. Only 16 percent of those relationships ended in separation. When women fell in love at first sight, more than 60 percent went on to wed that person, and only 7 percent of those marriages ended in divorce—a remarkable statistic given that, on average, nearly half of all marriages fail and end in divorce. Falling in love at first sight is a surprisingly successful strategy for selecting a partner.

With brain imaging, we can also see just how much romantic love is shaped by parental love. In figure 5, mothers were shown pictures of their children while they were in the fMRI scanner. Their brain imaging responses were then compared with the brain activity demonstrated when they were shown a picture of their spouse or romantic partner. What is striking is just how similar the activity levels are; they are close to identical. This is not to say that falling in love with your soulmate is like falling in love with your child. But these two potent emotions rely on the same regional activity in the brain. Parental love and romantic love occupy similar substrates on functional neuroimaging of the human brain. Again, it is another example of the scaffolding rule in action. It makes sense because, as children, the maternal bond (or its equivalent) is the first substantive experience of love we have. Everything that comes after that, including our sense of romantic love, is built on that foundation.

Sexual Preference and the Brain

Before tackling the profound implications of what brain imaging has revealed about sexual preference, identity, and orientation, it is helpful to ensure we are all comfortable with some of the recent insights into the nature of gender itself. During my education and training as a medical student and a resident, the notion of gender that was taught was that of an unchallenged, monolithic binary principle of chromosomal predetermination. However, decades later, science no longer embraces the notion that gender can be accurately represented as a purely binary concept. Instead, gender is more correctly described as a spectrum or continuum. This change in this position stems not only from clinical studies in humans but also a number of animal studies which confirm that gender is not restricted to simply male or female as defined by genotypic or phenotypic criteria. Variation in biological sex and gender expression has been well-documented across many species through scientific research. Such variation is observed at genetic, hormonal, and morphological levels. Models predict and experiments confirm that traits associated with sex and gender exist along a continuum rather than a strict binary. The lived experiences of gender diverse individuals further demonstrate that gender identity and expression are not solely determined by the genitalia with which one is born.

Numerous studies across various species, including humans, have found diversity in sexual traits, behaviors, and identities that challenge the simplistic male-female binary model. This diversity spans genetic factors like sex chromosomes, hormonal profiles, genital morphology, secondary sex characteristics, and brain structures related to sexual behavior and identity. The existence of intersex individuals who do not fit neatly into male or female categories provides clear evidence that sex is better represented as a continuum rather than two discrete categories.

We can look at current brain imaging and ask, how does attraction figure into our notions of sexual preference and orientation? For this kind of evaluation, test subjects are typically asked to watch videos while lying in an fMRI scanner. The subjects are shown erotically neutral videos of sporting events as a control, and at random intervals erotic film segments are interspersed. The scans revealed that sexual attraction was a complicated orchestration of responses of different brain areas.

Fetal gender differentiation begins halfway through gestation. The typical male XY chromosomal arrangement induces male characteristics only when adequate secretion of testosterone occurs. High levels of circulating testosterone give the brain its male anatomical configuration. When little or no testosterone circulates, the brain will go on to develop female characteristics. However, the impact of circulating testosterone can be affected by a host of factors. These include the density of receptors on target tissues, the presence of competing enzymatic reactions, and inborn errors of metabolism. One's genetic makeup, therefore, does not necessarily align itself with the functional anatomy of one's brain, one's sexual identity, or even one's external sexual characteristics—let alone one's gender.

Ivanka Savic and Per Lindström of the Stockholm Brain Institute used brain imaging studies to evaluate large groups of heterosexual and homosexual men and women. They found that *heterosexual men* and *homosexual women* had similar patterns of brain activity, particularly in the amygdala on the *right* side of the brain. By contrast, *homosexual men* and *heterosexual women* (both more attracted to males) showed activation of the amygdala on *both* sides of the brain (see figure 6). The takeaway message is not just what areas of the brain are activated but also the ability to look at patterns of regional activation to predict one's sexual orientation. We see that *heterosexual men and homosexual women, on the one hand, and homosexual men and heterosexual women, on the other, show similar patterns of brain activation in response to images of men and women.*

Sexual Identity

Recognizing the differences between sexual preference, orientation, and identity is critical. Sexual preference, of course, takes in a lot of territories: gender, dress, physical attributes, sexual practices, fetishes, temperament—you name it. Sexual orientation is considered a more stable and constant parameter of partner preference. Sexual identity finally refers to an internalized awareness and perception of one's own gender. One could term it an individual's sexual consciousness. Grouping by sexual orientation relates to heterosexuality, homosexuality, and bisexuality. However, increasingly, terminology such as "androphilia" and "gynephilia" is preferable because it is more inclusive of intersex and transgender individuals.

Gynephilia, Androphilia, and Bisexuality

One standard research method to assess an individual's sexual attraction is to measure subjects' genital arousal (as indicated by increased blood flow to the genitalia) while viewing erotic photos or videos. The norm is that heterosexual men demonstrate arousal when viewing women, while heterosexual women can show equal arousal when viewing either gender. Similarly, we see the anticipated differences in arousal in homosexual men and homosexual women. However, bisexuality offers a different perspective.

A research paper published in 2011 by the Kinsey Institute evaluated test subjects as they watched videos of heterosexual, bisexual, male homosexual, and lesbian activity while data was collected on genital and subjective arousal in men and women of different sexual orientations. Bisexual men and women showed no differences in arousal levels when viewing images of men than did homosexual men or heterosexual women. Similarly, bisexual subjects showed similar levels of arousal when viewing images of women as lesbian subjects and heterosexual men. Bisexuality, then, would seem to be the type O negative of sexual orientation—a distinct, universal sexual orientation. In the world of medicine, the O negative blood type is considered "the universal donor" because you can transfuse O negative blood into anyone who needs blood without an allergic reaction. Therefore, *bisexuality appears to represent a distinct, universal sexual orientation.*

Increasingly, there is recognition that sexual orientations and interests are not fixed for life. They can vary and fluctuate within the same individual over the short term or evolve throughout an individual's romantic life span. A recent study of more than thirty thousand individuals in the United States demonstrated that women in their twenties and thirties were three times more likely to report being bisexual than men. There was also a pronounced trend in older adults to come out later in midlife, often in the wake of adult children having left home or marital differences having been resolved. It is critical to acknowledge that *gender identity and sexual orientation do not lend themselves to a binary distribution nor are they fixed for life.*

The Transgender Brain

Sexual preference, orientation, and identity result from complex interactions of genetic and environmental influences. They can vary by genome, upbringing, physical habitus, hormone concentrations, receptor density, and differential activation of brain function, to name a few. Brain imaging is still in its relative infancy when it comes to deciphering human sexuality. That said, it has begun to make some inroads.

One of the most underrepresented groups to be assessed with brain imaging is transgender individuals. Such subjects often identify early in life that they perceive themselves as "men trapped in a woman's body" or vice versa. In Madrid, Spain, Antonio Guillamón and his research team reviewed the brain scans of eighteen female-to-male (FtM) transgender subjects *before* they ever initiated any cross-sex hormone treatment or surgery and compared them with a control group of twenty-four males and nineteen females. They analyzed the direction and complexity of white brain fiber tracts in these three groups. They found that the FtM subjects showed a white matter organization that most closely resembled male rather than female neuroanatomy. These results are compelling new evidence that structural differences in the brains of FtM transgender individuals exist long *before* they undergo any treatment as part of a transgender sex change.

Neuroimaging has begun to weigh in on the notion of gender dysmorphia (GD), which is the subjective discomfort or upset in individuals whose sense of gender identity differs from the one assigned to them at birth or is attributed to them based on physical characteristics. Much has been made recently of transgender clinics that attend to patients with GD as minors. However, recent fMRI studies have revealed that while gender-dysmorphic and gender-atypical changes in connectivity in the brain do not generally appear before children enter puberty, this issue usually manifests itself in the pediatric age-group and, certainly, while children are minors. The second trend becoming apparent from functional brain imaging is that transgender brains are more like their desired gender at an earlier age. This suggests that altered brain development precedes the onset of GD; therefore, transgender children may also experience issues surrounding gender development and sexual orientation at an earlier age than the general population.

It lends substantial evidence to the claims voiced by so many transgender individuals that they feel as if they were trapped inside bodies of the wrong gender. Their brains are driving them to find the gender framed by their neuroanatomy. This means gender is a matter of neuroanatomy, and not only simple, personal preference.

The furor raised over transgender people gaining access to bathroom facilities compatible with their gender identity is ridiculous for two reasons. The first is the neuroimaging data confirms that the sexual identity selected by transgender individuals reflects their underlying neuroanatomy. Period. The second is practical: the issues of transgender use of toilets can be inexpensively resolved. Just offer individual unisex toilet facilities that can be locked for privacy. Done. Finally, let's look at the facts. The truth is when law enforcement organizations across the country were queried about problems arising from transgender use of toilet facilities according to sexual identity, not a single agency could find a report of even one incident on record.

All You Need Is Love

An update of the most comprehensive longitudinal research study in history sheds light on our need to create and sustain loving relationships. The Harvard Study of Adult Development has been following two cohorts of young men: an initial group of 268 men recruited from the sophomore classes at Harvard College between 1939 and 1944; and a second cohort of 456 young men recruited from the poorest, most disadvantaged neighborhoods in Boston during the same interval. Both groups were followed prospectively. At the time, no woman was enrolled as a subject in the research.

The Harvard study asked, what factors can be identified that make these men feel happy and successful? Subjects were interviewed at regular intervals during their lives and could be queried regarding their goals at various points during their lives. The study also collected surveys of career satisfaction, marital stability, and retirement satisfaction. It included physical and mental health evaluations, review of medical records, blood tests, and brain scans, as well as inventories of intelligence, memory, and cognitive functions. The data also included interviews with spouses and children. Most recently, a new generation that comprises more than two thousand individuals who were the offspring of the original subjects has been included for follow-up study.

In 2012, George Vaillant published the interim results when the original subjects were between their mid-sixties and their mid-seventies. His first finding was that financial success (which the subjects had considered an important goal early in their lives) was related to *the affection experienced in social relationships.* It did not correlate strongly with intelligence or one's initial socioeconomic status. The overarching finding of the study was social relationships—the sense the subjects had of being closely connected to the other people in their lives—were the single most important predictor of physical and mental health and longevity. Vaillant summed it up succinctly: "Happiness is love. Full stop."

Human and animal data highlight the profound role loving relationships play in our development. The template of the mother-infant relationship provides the foundational wiring diagram of how the human brain organizes and prioritizes relationships. These studies also show the brain tolerates very little when it comes to developmental shortcuts. Neuroimaging reveals a distinct beauty in how the brain frames love and sexuality. The studies demonstrate that the love felt by a young married couple can be truly undiminished even after fifty years in a relationship. They highlighted how the love and warmth provided by our parents and families in our seminal relationships kindle our ability to express affection in our lives.

The brain has marshaled an abundance of resources, functional systems, and powerful neurotransmitters to bolster this supreme sensation of love. This commitment of neuronal and neurochemical assets speaks to how vitally important it is to us as a species and as individuals. It is clear that without love the functional development of the human brain is disrupted—if not outright broken—and unable to reach its full potential. *Without love in early childhood, success becomes far less attainable.* Every relationship we make is predicated upon it. The latest studies that have looked at love tell us human purpose flounders and risks being irretrievably lost without it.

4

On the Nature of Empathy, Altruism, and Kindness

Three things in human life are important. The first is to be kind.
The second is to be kind. And the third is to be kind.
—HENRY JAMES

A Killing and a Kindness

One winter night, a fourteen-year-old boy roamed the streets of Baltimore. He had no home and no family. The only individuals who would occasionally offer him a place to sleep or a bite to eat were members of a notorious street gang that was heavily involved in dealing drugs and street crime. While it was a world of violence, the boy wanted to belong to the gang. To be a full-fledged member, he would have to fulfill one requirement: he had to murder someone. That was why he wandered the streets that night with a pistol stuck inside his belt.

As he walked down one darkened street, he could see a teenager walking toward him. The young man appeared to be alone, and no one else was in sight. He saw his chance. He pulled out his gun and fired several shots at point-blank range into the stranger. The victim died moments later. Now he had proven to the gang he was "bad" enough to earn full membership status. The following day, the murder was front-page headlines.

But instead of welcoming him into their midst, the gang members turned him over to the police. The murder was attracting too much scrutiny.

The slain teenager had been only three years older than his killer. His murderer had never set eyes on him before he shot him. Never uttered a word to him. He did not even know his victim's name until he heard it when he was being arraigned in court and charged with first-degree murder. The victim had been the only child of a single mother.

Once the murder trial began, the defendant sat quietly alongside his attorney. The grieving mother attended the trial every day and was always in the same chair in the courtroom. Her unflinching presence unnerved the defendant. Day after day, the woman stared impassively at him. She never spoke a word—until the last day of the trial.

The killer was still a minor, and he was sentenced to serve out three years until he reached the age of eighteen. As the bailiff led the convicted teenager out of the room past her, the woman stood up. She identified herself as the mother of the murder victim. She then looked the defendant right in the eyes and, shaking her finger at him, announced defiantly, "I am going to kill you!" The young man gulped and followed the bailiff out in handcuffs.

Almost a year had passed since the trial. No one ever visited the teenager in prison. No family. No friends. No gang members. One day a letter arrived at the penitentiary addressed to him. It was from his victim's mother. She was asking if she could visit him in prison. Inside the letter were a few dollar bills so he could buy some cigarettes and candy at the commissary. Although he dreaded her visit, he also ached for company. Every other prisoner had someone visit, someone with whom they could talk. He had no one. He finally gave her permission to come.

When she finally came on visiting day, she was full of questions. She asked him about where he grew up, how he was raised, about his life on the streets, and, finally, about why he had killed her son. Then she left him a carton of cigarettes. As she handed over the smokes, she asked if she could write to him and visit again, perhaps. Eventually, she got into a routine so that a new letter would arrive for him at the prison every week. Besides her, no one wrote to him or visited.

As the relationship deepened, she persuaded the boy to take classes with the tutors in prison and get his high school equivalency diploma. He did, and he got good grades for the first time in his life. Soon his release date began to loom on the horizon. When the mother asked him what his plans were, he told her he had none. He doubted anyone would even give him a job, knowing he had just gotten out of jail for murder. She eventually arranged to have the young man hired by a firm run by a friend of hers. Where would he live once he got out? No idea. No family. No friends. Probably back on the street. She considered his dilemma for a while, and then she suggested, why not stay temporarily with her since she had a spare room?

When he was released, the young man began working, and for the next eight months he stayed in the house where his victim had been raised. He ate food at the same kitchen table and slept in the same bed. And he spent hours talking with his victim's mama.

One day she asked him to come into the living room so she could talk with him for a few minutes:

"Do you remember in the courtroom when I said I was going to kill you?" she asked.

"I sure do," he replied. "I'll never forget that moment."

"Well, I did [kill him]," she went on. "I did not want the boy who could kill my son for no reason to remain alive on this earth. I wanted him to die. That's why I started to visit you and bring you things. That's why I got you the job and let you live here in my house. That's how I set about changing you. And that old boy, he's gone. So now I want to ask you, since my son is gone, and that killer is gone, if you'll stay here. I've got room, and I'd like to adopt you if you let me."

Eventually, they did just that. They became the family to each other that circumstances had taken from both. How can human beings have this capacity to care for each other and forgive each other?

Impulsive Altruism

Acts of altruism can be impulsive. When we are stressed, our frontal lobes become inefficient, so we literally can find ourselves acting out of emotion. During a 2015 terrorist attack in a Paris café, a bystander named Ludovic Boumbas saw a terrorist aim at a woman as she crouched in the corner for cover. Boumbas dove on top of the young woman and was killed by the bullet intended for her.

Acts of altruism tend to be acts of emotional compulsion that can include sacrificing one's own life on behalf of a stranger. So, why didn't Boumbas's brain tell him to save his own life first? To understand such acts of altruism, David Rand at Yale University did a series of studies where individual subjects were asked to play a game for money. He found that when he rushed and confused players by asking them to memorize numbers simultaneously as they played, the frazzled test subjects were more likely to share their cash with their fellow players. In other words, the less time individuals had to think, the more likely they were to be generous.

In the last chapter, we saw that heroism is rooted in love, the most potent feeling we have. And, as an emotion, courage arises from impulse, not reasoning. Rand evaluated more than fifty individuals who had been awarded the Carnegie Medal by the Carnegie Hero Fund. Andrew Carnegie personally established the medal in 1904 to recognize acts of heroism in civilian life. The Florida resident Charles Carbonell Sr. was awarded a Carnegie Medal in 2007 for coming to the aid of a policeman who was struggling with an assailant trying to wrest control of the officer's revolver. Carbonell won a second Carnegie Medal in 2014 for pulling a woman from a burning car wreck in 2011. After being recognized twice for his courage, he admitted he didn't have time to think when he acted heroically. "If I'd thought about it first, I probably wouldn't have done it," he admitted. "I always do it first and then think after." And that is how it works: valor is acting on one's impulse before the frontal lobes put a stop to it. I'm sure Boumbas would have said the same thing.

Altruism in Nonhuman Species

Altruism is not limited to humans. Animals may practice it better than we do. And they cannot deduct acts of philanthropy on their taxes or get plaques at the annual charity ball. In 1959, Russell Church, a researcher at Brown University, published a provocative article titled "Emotional Reactions of Rats to the Pain of Others" in the *Journal of Comparative and Physiological Psychology.* Church had taught rats to pull a lever to obtain a food pellet. A simple enough trick. Then he set up another rat in a neighboring cage. This time, when the trained rat pushed down on the lever to get its food pellet, it also sent an electric shock coursing through the floor of its neighbor's cage. And it gave that neighboring rat quite a jolt. Naturally, the neighboring rat shrieked in pain and then cringed in the corner. Very quickly, however, the first rat figured out that pressing down on the lever was responsible for producing the painful shocks that tortured its next-door friend. The result was that the trained rat stopped requesting any food (the lever gave it access to its sole source of nourishment). The rat refused to push the lever—even to the point of starving itself. In other words, a rat was willing to starve itself to death if it meant he would not inflict suffering on a fellow rat. Hmmm. I wonder how humans would fare on a similar test.

What about altruism displayed for members of a different species? Such a demonstration happened by accident on August 16, 1996, when a three-year-old boy climbed over the fence of the gorilla enclosure at the Brookfield Zoo in Illinois and fell eighteen feet down into the heart of the gorilla compound. Binti Jua, an eight-year-old female gorilla who had been raised by humans at the zoo, took the unconscious boy in her arms and carried him to safety so an attendant could tend to the child.

Similarly, in 2008, a different kind of incident was recorded on video off the coast of New Zealand. In this footage, one can see a bottle-nosed dolphin come to the rescue of two beached whales, a mother and her calf. The whales had gotten themselves grounded and could not figure out how to get to deeper water because a large sandbar blocked their way. The dolphin came up to the whales and then showed them a two-hundred-meter-long detour around the bar that would lead them directly to open, deeper water. These examples of cross-species altruistic behavior are significant because *no issues of genetic preservation or selection motivate such acts of compassion or sacrifice.* Now the behavior of the mother in Baltimore does not seem so bizarre. We seem to come full circle: altruism seems to occur because it is its own reward.

The Mirror Neuron System

The Discovery of the Mirror Neuron System

Giacomo Rizzolatti and Vittorio Gallese at the University of Parma carried out research for decades with brain electrodes in monkeys. By the early 1980s, they had begun advancing thin platinum wires into a specific area labeled F5. The F5 area is located in the inferior frontal lobe of macaque monkeys, and it was targeted to study how motor tasks were organized and initiated in the brain. But, like in so many watershed events in the field of science, the truly important discovery came about as a result of serendipity.

The schedule of lab experiments demanded that the monkeys with the implanted electrode grids be hooked up to a machine that recorded the firing of their cortical neurons. The experiments were run in the morning. Once the morning's data was collected, the recording equipment was turned off. After lunch, the recording devices were started up again and a second round of experiments was initiated. For this particular "run" of experiments, the researchers were evaluating how the monkeys organized their motor movements when they reached out for a peanut.

On this one particular day, everything was running late. When everyone broke for lunch, there was less time than usual to eat. Now, remember this was in Italy, where cuisine and the dining table's camaraderie are considered essential to the rhythm and meaning of life. The head lab technician decided he could save time by leaving the monkeys hooked up to the recording equipment rather than tearing everything down and having to reconnect and recalibrate it again after lunch. It would also save him some valuable time for lunch. He grabbed a gelato to go and headed back to the lab. While waiting for the senior researchers to return, he licked away at his gelato on its cone.

The research team showed up shortly thereafter. The scientists looked down at the tracings on the recording machine and asked the lab tech what he had been doing to the monkeys during the break. Completely innocent, he replied, "Nothing. I just waited for you while I ate my gelato." The researchers were dumbfounded. There was activity all over the place. The technician swore he hadn't touched the monkeys. But as soon as he went back to licking his gelato, the monkeys' cortical neurons started firing like crazy. The recorded activity made it look as if the monkeys were eating the gelato themselves. Something out of the ordinary was happening.

The next day the team set about recording again from F5. This time, they went back to their peanut protocol. They offered the macaques a peanut and looked at the cortical firing pattern. Then they had the lab tech pick up a peanut. Neurons in the F5 field showed the same firing pattern they had recorded when the monkeys grabbed the peanut for themselves. Rizzolatti would later comment, "We were lucky, because there was no way to know such neurons existed. . . . But we were in the right area to find them."

This group of neurons in the F5 area of the primate cortex seemed to function as "mirrors." These brain cells fired whether the monkey carried out the activity itself or merely witnessed it being carried out by the tech.

"Mirror neurons," as they became known, had other properties. For example, the monkey did not even have to see the peanut being picked

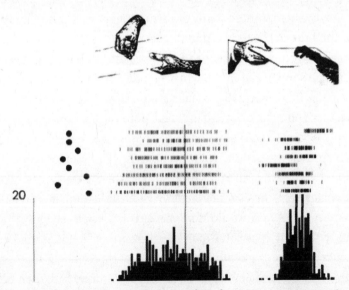

Mirror neuron activity is seen with the monkey observing someone grasping the peanut (*upper left-hand panel*) versus the monkey grasping the peanut itself (*upper right-hand panel*). Beneath that we see mirror neuron activity in both experimental conditions. The amount of mirror neuron activity is almost identical whether the monkey grasps the peanut or is watching someone else grasp the peanut.

up. If the monkey heard the peanut being crunched, then the mirror neurons would fire. A recording of a peanut being crunched caused the same action. It was the same result if the task was switched to watching a technician rip a piece of paper, having the macaque rip the piece of paper himself, or having the monkey just listen to a recording of a piece of paper being torn.

This new class of neurons offered exciting clues as to how we might learn through mimicry, but soon it also provided a new mechanism as to how one individual could gain insight into the experiences and actions of another. In 1992, Rizzolatti and Gallese wrote up their results and submitted them to one of the most prestigious scientific journals in the world, *Nature*. The journal editors politely rejected the material for publication because of what they cited as a "lack of general interest." It was later published in a much less well-known journal, *Experimental Brain Research*. A decade later, the neuropsychologist Dr. V. S. Ramachandran would call the discovery of the mirror neuron system "one of the 'single most important unpublicized stories of the decade.' "

Discovering mirror neurons turned out to be the easy part. Trying to figure out why there were so many of them interspersed throughout the primate and human cortex was not. It seemed confusing—as if the brain did not care if you were the actor or the spectator. It was assumed that there must be some value in allowing the brain to mimic the actions it was observing, like a skier who visualizes himself going through the gates on the slalom run before he actually steps onto the course.

It soon became apparent that mirror neurons were not just for motor (or muscle) function. There were sensory mirror neurons, too. If I watched you burn your right index finger, then the area of my own sensory cortex that corresponded to my right index finger would also show activity. My brain would demonstrate activity to suggest that it understood the pain that you experienced by processing it as if it could have occurred to me. Later studies showed that there were also areas of the brain that mirrored emotion. If I saw that you were sad, then emotional areas of my own brain would fire as if my brain understood what you were feeling by mirroring your emotions.

An even broader picture of the mirror neuron system emerged. If empathy could be understood as the process of seeing the world through

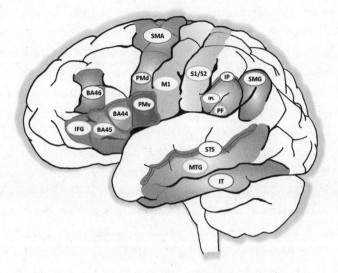

The human mirror neuron system is a relatively recently
discovered network that is located widely throughout the
cortex of the human brain. All of the labeled areas in this figure
indicate cortical regions where mirror neuronal activity has
now been demonstrated.

someone else's eyes, then here was a brain mechanism enabling the individual to process the sensory, motor, and emotional experiences of others
by experiencing them for him- or herself through mirror neurons. *The
mirror neuron system allows us to simulate at the neuronal and regional
levels in our brain what other individuals are experiencing in theirs.*

Brain imaging has revealed a high density of mirror neurons in an
area of the brain that lies adjacent to the prefrontal cortex and is called
the insula. The insula is the site of some complex emotional reactivity.
For example, it is the area where we try to interpret the facial expressions
of others. It is also involved in *interoceptive awareness*—namely, the ability to sense what is happening inside our bodies. For example, how fast
are our hearts beating or how distended do our stomachs or bladders
feel? The insula is also the site where there appears to be a convergence
between how we feel in a visceral sense (for instance, "this makes me feel
sick to my stomach") and how we feel emotionally (for example, "you
make me sick"). It seems a natural place where the brain can gauge what
we are feeling and reference it to what others are feeling and experiencing.

There are also fiber tracts reaching from the insula to areas like the hypothalamus that exert substantial control over our bodily functions. This appears to offer a mechanism whereby our emotions produce powerful physical reactions. So, for example, we can suffer from so much anxiety at a public performance that we throw up, and the signals for such responses would be carried by these white matter fiber tracts that connect the insula and the hypothalamus.

Another critical area in the mirror neuron system is the anterior cingulate area. This area, more than any other, lends what I would call "emotional color" to what we see. For example, we might see what we recognize clearly as a toy baby doll getting knocked off a table and falling to the ground. Our reaction would be far different in terms of our emotional response if we saw it was a real baby. Even though the images may seem similar in content, the emotional value lent to one is far more potent than the other.

The anterior cingulate area, prefrontal cortex, and insula combine as parts of the mirror neuron system to help us generate our emotional assessment. In the brain, context changes content. The last structure involved in emotional processing is our old friend the amygdala, which helps provide the "gas pedal" to our emotional reactivity. It is the engine of fight or flight, the seat from which fear, rage, anger—almost all of our powerful and overwhelming emotions—arise.

The Compassion-Empathy Circuit

The linkage of the mirror neuron system to the limbic system has led scientists to propose that these components could constitute the equivalent of a "moral-ethical" circuit of the brain. The circuit would determine how we react to others and whether we feel the need to connect with, or distance ourselves from, them. The circuit would help us assess the motivations of others and whether their actions are intended to help or hurt us. It could explain why the components of the system might make some people more empathetic than others.

One method that researchers employ to better understand empathy and self-sacrifice is to ask volunteer test subjects to play a computer game called *The Prisoner's Dilemma*. It is a cooperative game and a favorite among neuroscientists because a test subject can patiently play the game on a computer while his or her brain is being scanned. In *The Prisoner's Dilemma,* the players interact as characters who are prisoners of war held in an enemy camp. The characters are placed in situations where they might (or might not) mutually cooperate to advance the prisoners' interests or, alternatively, could betray each other to promote their own individual needs. Players get points for resisting the game's camp guards who are always interrogating prisoners to obtain confessions or information about fellow prisoners. A behavior like saving a scrap of food for a fellow prisoner being placed on starvation rations would earn the player points. But cooperating with the guards can also earn you points while taking some away from the other prisoners. What happens if all the prisoners unite and resist together? That kind of teamwork would earn players the most points.

Brain imaging studies revealed that dopaminergically driven reward sites in the brain increased their activity when players worked to help their fellow prisoners. Players who sacrificed for their prison-mates felt a pleasurable reward.

An interesting intervention was planned halfway through one set of experiments using *The Prisoner's Dilemma.* All of the test subjects were assembled in a room. The investigators informed the players that someone in their group was cheating and was secretly cooperating with the guards in the game. In reality, no one was cheating, but the researchers wanted to see what would happen if the prisoners felt there was a traitor in their midst they could not trust.

As you might imagine, the idea that a fellow prisoner was cheating and taking advantage of the sacrifices of others while avoiding the punishments doled out by the guards raised a great deal of consternation among the players. Some of the subjects suggested that a kind of "kangaroo-style investigation" be launched and "the snitch" or "the traitor"—as they called the unknown prisoner—be rooted out. Others voiced concerns that such efforts might undermine the cooperation among the rest of the players who were playing by the rules. However, by far, the most interesting finding was from the subsequent brain imaging studies after doubts had been planted in the minds of the volunteers.

In about half the subjects, activity in their reward centers was *undi-*

minished; that is, the actions of one "bad apple" were not going to under-mine the satisfaction they derived from helping their fellow prisoners. But that was true for only half of the players. The remainder of the players were coaxed to continue playing the game, but their pleasure centers were no longer activated by their efforts. The reward for sacrifice had been ruined once they felt someone was secretly taking advantage of them—a lesson in the fragility of charity. *Altruism, empathy, and generosity are all rewarded by dopaminergic centers in the brain.*

Empathy, Doctors, and Patients

During the COVID-19 pandemic, frontline health-care providers were being exposed to scores of coronavirus patients and a score of deaths in a single work shift. This was an excellent time to ask, could empathy cause fatigue? Could it break down under incessant demand and unrelenting need? It turns out chronic exposure to pain puts clinicians at risk for "empathy fatigue." In a laboratory study conducted by the University of Chicago, medical residents and age-matched nonphysician controls were asked to watch videos of a human hand. In one scenario, they observed a hypodermic needle being introduced through the skin and into the hand. In another, they watched the skin being stroked with a Q-tip.

Each subject had their brain waves recorded on EEG. In particular, the researchers were tracking what are called event-related potentials. Event-related potentials are unique, identifiable brain waves linked to a specific stimulus—in this case, watching the video when the needle (painful stimulus) or Q-tip (non-painful stimulus) made contact with the skin. There were no differences observed in the potentials between the resident and the control groups with respect to watching the video of the painful stimulus. This was true when the residents first started their training. However, the further the residents proceeded into their training, the more the doctors' event-related potentials began to shrink. In other words, their brains became inured to seeing painful procedures. So, one response when we see too much pain repetitively is to shut it out, to stop ourselves from feeling it. *Constant, repetitive exposure to the pain of others leads to empathy fatigue.*

The Pain of Others Is Our Own

There are other prices to pay for feeling the pain of others. In one study, subjects were tested with an electrode pasted to their skin. The electrode would gradually heat up in discrete, precise changes in temperature. In this way, researchers could accurately measure when the electrode became hot enough that the subjects felt it had grown too painful.

As part of the study, the subjects were asked to watch video clips. A few were of subject matter considered emotionally neutral (like a cityscape), and the subjects were asked to rank their pain thresholds while the researchers carefully calibrated the temperature readings. Then the subjects were divided into two groups. One group was shown a video in which an individual (actually an actor) was being interviewed after he supposedly had just lost someone close to him. The scenario was aimed at evoking the test subjects' sympathy and compassion as they watched. The second group watched a video with the same actor being interviewed, only this time the subjects were told that he had just been arrested for stealing. It took less heat to make a test subject feel pain if they felt compassion toward the individual depicted in the video than if they did not. In short, *the more the subjects felt empathy for the pain of others, the more susceptible they were to experiencing pain themselves.*

Our planet is still recovering from a global pandemic, one of the most destructive the planet has seen in more than a century. The worst ever was the Black Death that spread across Europe from 1346 to 1353. In that period, 50 million people perished. That represented 60 percent of the entire continent's inhabitants. In 1918, a deadly H1N1 virus—it became known as the Spanish flu—swept around the globe and infected more than 500 million people, killing an estimated 50 million. The first case in the United States occurred in an army cook at Fort Riley, Kansas. The cause of his mysterious illness was correctly diagnosed in Haskell County, Kansas, by a local doctor named Loring Miner. Dr. Miner felt compelled to issue a written warning in nothing less than the journal of the U.S. Public Health Service to alert authorities about the presence of this perilous index case. Dr. Miner's warnings went unheeded. A week later there were 522 new influenza cases at the fort. One-third of the planet's population would eventually perish in the 1918 pandemic.

If there has been one recurrent weakness in a century's worth of pandemic experience (that includes the COVID-19 crisis), it is our lack of insight that the calamity that now besets others will soon become our own. The real danger lies not in what we feel but in what we *don't*. Pandemics reinforce the lesson that the failure to be moved by empathy leads to inaction and that, in turn, translates into lives needlessly lost—including those of people we love or our own. The capacity to feel pain—as overwhelming as that can be at times—is preferable to remaining numb.

5

On the Nature of Psychopathic Killers and Mass Murderers

After a while, murder is not just a crime of lust or violence. It becomes
possession. They are part of you . . . [the victim] becomes a part of
you, and you [two] are forever one . . . and the grounds where you
kill them or leave them become sacred to you, and you will always be
drawn back to them.
—TED BUNDY, convicted serial killer, kidnapper,
   and rapist of more than thirty women

Quantifying Evil

Neuroimaging studies have provided a clearer picture of the many
moving parts of the moral-empathy ecosystem within the brain.
This circuit is not just part of the fabric of our self-awareness,
but it contributes to our codes of conduct about morality, fairness, and
justice. This circuit can give us loving, supportive parents or provide the
world with a Mahatma Gandhi. But a malfunction of this same empathy
circuit is implicated in everything ranging from mild forms of autism at
one extreme to outright psychopathy at the other.* The balance between
good and evil in us is a measure of the health of our empathy circuit.

Scientists have developed specific inventories and assessments to vali-
date the workings of the brain's moral-empathy circuitry, but they also

---

* I do not mean to imply that all forms of autism relate purely to dysfunction of the empathy circuit,
but a great deal of research data suggests that some forms or some aspects of autism may be related
to dysfunction or delayed development of areas in the empathy circuit.

point out it can be primed by early childhood experiences to malfunction.

Dr. Simon Baron-Cohen is a psychologist who focuses on such malfunctions. As a child, he remembers being affected by the stories his parents told about the Holocaust. One story haunted him: his father told him about an evil experiment where Nazi surgeons removed a woman's hands and reattached the right hand to the left wrist and the left to the right wrist. The doctors wanted to see how well her brain could sort out the inversion. The answer is that it cannot, and such sadistic surgery renders both hands completely and permanently useless. That image stayed with Baron-Cohen all his life.

Later, as an adult, Baron-Cohen researched how to codify or stratify human empathetic responsiveness in his book *The Science of Evil: On Empathy and the Origins of Cruelty.* He developed the Empathy Quotient, and he collected data to allow the quotient to stratify empathetic responses into seven different levels.

Individuals in Level 1 lack self-restraint, and they hurt people because they cannot curb their own violent, aggressive impulses. Level 1 individuals are capable of hurting someone and then, later, reflecting cognitively on the event. They are swept away in the moment by their emotions, like anger or jealousy, that remove the cognitive inhibition that kept the individual from acting on his impulses. An example might be someone who beats an individual up because he believes that person insulted him by stepping ahead of him in a queue.

The emotions of individuals at Level 2 are affected by outside influences. Yes, they can still hurt people, but they can also feel regret over their actions. But not enough to make them stop hurting people. And while they can be externally motivated to act in a particular way, they fail to do so out of any intrinsic compulsion. So, for example, someone might say, "You should help that old lady with her bundles as she crosses the street." They will do what they are told but feel no intrinsic urge to do so.

Individuals in Level 3 still suffer from a substantial empathetic disability. Level 3 individuals must constantly remind themselves to act "normally." It is almost as though they must read a script to tell them how to act. They have been repeatedly instructed how they are supposed to behave, but they have to talk themselves into playing the part. A good example would be an older sibling who knows he's not supposed to thrash his younger baby sister just because she picked up a toy that did not belong to her. But it would not be his natural impulse to overlook her youth and lack of insight. Level 3 individuals have a hard time socializing because they cannot feel the empathy that others do.

Level 4 individuals exhibit average to below-average empathy. Individuals in Level 4 are uncomfortable having to confront or deal with the emotions of others. They prefer to avoid immersing themselves in deeply emotional encounters and like the conversation to stay superficial (for example, sports). Unfortunately, a substantial number of men fall into this level. Level 4 individuals do not feel genuine emotional attachments but rather learn to develop relationships based on functional convenience. So, Level 4 individuals might form a friendship with someone working the same factory shift not because they like the person but rather because they're together so much of the time on the shop floor. Far more men test out to be Level 4s than women.

Conversely, in Level 5, females outnumber males. Relationships are based on mutual respect, emotional need, and rapport. These are individuals who, for example, might listen to their friend talk about a crisis and try to give them sincere advice. They exhibit a genuine interest in eliciting and understanding the feelings of others and sharing those feelings.

Only a small minority of individuals in the population will test out high enough to qualify for Level 6. We would refer to them as empaths. These are those rare people who can put their agendas aside for the sake of helping others. Many people who end up in careers as therapists, nurses, doctors, or counselors will often test out in this level.

However, two additional categories are missing from this original schema that Baron-Cohen developed. He revised his earlier hierarchy and added Levels Zero Negative and Zero Positive.

Level Zero Negative: The best way to describe "zero negative" is that the category was meant to include individuals who, in a sense, have fallen over humanity's edge into the abyss. Individuals in this level include severe psychopaths, extreme narcissists, and, often, outright sadists. These are individuals who are capable of absolute and methodical dehumanization of their victims. They are incapable of feeling guilt, remorse, or repentance. An example might be a serial killer who keeps his victims alive so he can torture them longer.

Zero-negative individuals exhibit some unusual characteristics. For example, imagine you placed a zero-negative individual in a brain scanner while you flashed an image on the computer screen in front of their eyes. The image is one of a hand with a bleeding cut across the palm. Now, while they are staring at this image, you instruct them to imagine it is their own hand. You will see no difference between their brain activity and a normal control group on the fMRI scans.

By contrast, when you ask the same zero-negative individual to imagine that they are looking at a photo of someone else's hand—not their own—you will see a dramatic difference. There will be widespread activation of the brain's mirror neuron system in the normal control group. The reaction you will see on the fMRI scan will not be very different from the activity you saw when these volunteers imagined the injury to their own hands. In short, their empathy levels are high. But, in zero-negative individuals, when you ask them in the scanner to imagine the injury happened to someone else, there is no response. There is no activation of the mirror neuron system: they simply cannot fathom the suffering of others.

What happens in the brain to make a zero-negative individual? First, imaging studies now show us that there is a small subgroup of zero-negative individuals who are born with genetic mutations of their oxytocin receptors. The result of the mutation is their brains are effectively unaffected by circulating oxytocin, and they are functionally oxytocin deficient. These individuals, therefore, experience a lack of bonding, closeness, and trust with their fellow men and women. They are disconnected from people. For this reason, in neuroimaging studies, they exhibit less empathy, more antisocial behavior, and increased aggression toward others. This means that individuals classified as zero negative (on the empathy scale devised by Baron-Cohen) exhibit deficiencies in activation of the mirror neuron system.

Imaging studies of zero-negative individuals have also demonstrated that impaired frontal lobe function can lead to more significant emotional contagion. The ability to suppress or downregulate emotions correlates positively with one's capacity to feel concern for others. The ability to recognize distress cues in others also serves as a potent inhibitor of violent behavior. Both these mechanisms are impaired in individuals with severe empathic dysfunction.

Baron-Cohen also identified another level: Zero Positive. It was a much more controversial classification. For decades, Baron-Cohen had been researching autism spectrum disorder. There is a correlation between the increasing severity of autism and decreasing activity in the mirror neuron system on brain scans. Baron-Cohen hypothesized that high-functioning individuals with autism spectrum disorder (for example, Asperger's syndrome) are polite in social situations because they have been taught to be polite—not because they feel intrinsically respectful of others. It may seem harsh to place autism spectrum disorder into this empathetic framework. Still, it offers an intriguing way to look at some (*not all*) individuals with autism; namely, as a syndrome of underdevelopment of the mirror neuron system. Many individuals diagnosed with autism early in childhood can improve dramatically over time because early intervention takes advantage of the brain's plasticity to help address or reinforce the deficiencies in the mirror neuron system. Clinical trials are under way with the administration of exogenous oxytocin in autistic children and show great promise.

A Scientist Hunts Himself Down

Jim Fallon is a neuroscientist who has been studying the biology of psychopathic killers for decades, looking at everything from their DNA to their brain scans. In 2006, his research took on a very personal nature. He had been at a family picnic where his then eighty-eight-year-old mother mentioned in an offhand remark that, given his research interests in murderers, *he ought to look into his own family tree!* His mother remarked that a few members on his father's side of the family were "cuckoos"—with histories of violence. What he found was unsettling: stretching as far back as his great-grandfather, who had been hanged for murdering his mother, the family line had produced no fewer than seven murderers. Among them was the infamous Lizzie Borden, a New England woman who was strongly suspected of having murdered her father and stepmother by crushing their skulls multiple times with an ax.

As Fallon was uncovering his family background, new brain imaging research was emerging on psychopathic criminals. It showed they had profoundly decreased activity in the prefrontal cortex and the temporal

A photograph of the infamous Lizzy Borden.

lobes where the amygdala is located. The fiber bundles connecting the two lobes—called the uncinate fasciculus—were also strikingly diminished in size. It began to make some sense. The decrease in amygdala activity helped to explain why so many serial killers never seemed fearful or concerned that the police would find them or that they would get caught. The decrease in prefrontal cortex activity also meant there was little or no restraint on their emotional behavior. And, finally, the same pathway, the uncinate fasciculus, that was supposed to put the "brakes" on their impulsive behavior was small and underdeveloped in size. Brain imaging studies could help explain some of the behaviors exhibited by psychopathic criminals.

What was emerging from brain imaging studies was that the brains of serial killers showed unique features that helped explain their ability to carry out horrific acts of violence without pity for their victims, fear of being apprehended, or pangs of conscience.

At first, Fallon was shocked to find evidence of a possible trait that predisposed members of his own family to violent behavior. Earlier in his career, he had been working on an unrelated project on Alzheimer's dementia and, fortuitously, had included nearly a dozen of his close relatives in the study as a part of a standard control group. This meant that he had both their blood samples and their PET scans in his database. Looking up the archived brain imaging studies, he could see that all of the imaging studies of his family were normal—except for one: his own. On his own scans (see figure 7), Fallon saw dark regions of low activity in the prefrontal cortex and the temporal lobe, along with an underdeveloped uncinate fasciculus. His scan fit the classical neuroimaging findings seen in psychopathic killers.

Brains and the "Warrior" Gene

Given the PET scan data, Fallon wanted to dig deeper. With the blood samples he had stored from the Alzheimer's project, he could request a genetic analysis of his family members' DNA be performed. There was one particular gene he wanted to know about: monoamine oxidase A (MAOA). The MAOA gene was first discovered in an unusually aggressive, very territorial breed of lab rats. Shortly after that discovery was made, a similar gene was found in the human genome. Individuals whose DNA tested positive for the MAOA gene tended to behave more aggressively and violently than individuals who did not have the gene. There was a higher incidence of the gene in prison populations. The announcement in the scientific community that there might be a genetic marker for violent behavior made newspaper headlines. Controversy sprang up: Should juvenile delinquents and criminals be screened for the gene? Police officers? Ethnic studies sprang up, and the same gene was also identified among Masai and Maori tribesmen (the gene is found exclusively in males), two cultures renowned for their traditional warrior codes. The popular press nicknamed it the warrior gene and the name stuck.

MAOs are complex enzymes located throughout the body. Their job is to clip a small chemical group, called an amine, off larger molecules to inactivate them. The family of molecules with amine groups in the brain includes two of our now familiar neurotransmitters, serotonin and dopamine, and a third one called norepinephrine. Levels of serotonin circulating in the blood and the brain are closely associated with mood, while dopamine can often motivate individuals to act aggressively. Norepinephrine can prime the brain for violence.

When MAOs function in their usual fashion and operate at high efficiency, they inactivate these powerful amine neurotransmitters. MAOs act like great white sharks, cruising the synaptic spaces between brain cells, seeking out amines, and then chomping down on them and crunching off a group. They leave behind a crippled molecule that can no longer function. If there are inadequate levels of the predatory MAOs circulating, then excessive levels of amines accumulate. Unfortunately, many mammals—from mice to men—exhibit violent, aggressive behavior when amine concentrations get too high. A mutation in the MAOA gene makes the enzyme less efficient at inactivating these amine neurotransmitters and is associated with increased tendency toward aggressive behavior.

The result is a brain that is chemically "primed" to react and show aggression.

Fallon needed to know if he or any other men in the family had the MAOA mutation. He sent the stored blood samples off for DNA analysis. No men in the family exhibited the dreaded MAOA gene—except for himself. Jim Fallon is a preeminent scientist and researcher. He is a smart, thoughtful man and, in real life, does not vaguely fit the profile of a psychopath. But on paper, he has all the right features. As Fallon himself once put it in an interview, "You see that? I'm 100 percent. I have the pattern, the risky pattern. . . . In a sense, I'm a born killer." Fallon isn't. Nor is his wife concerned, pointing out she has known her husband since they were both twelve years old.

Genetics is a weird science. Genes may help us identify tendencies and proclivities, but they cannot predict the future.* One crucial factor is how the environment affects genetic susceptibility. Individuals may have the genomic markers, but if their environment is loving, supportive, and nurturing, then that proclivity becomes muted or suppressed. We now understand through the epigenome how our home environment, culture, parenting, schooling, nutrition, and even our formative experiences can directly influence whether inherited gene sequences in our DNA are energized or muted. Clearly, in Jim Fallon's case, his environment

---

* There are a handful of lethal gene mutations where the stricken individual is doomed to die (for example, Huntington's disease and Niemann-Pick disease).

lent him an epigenome that safely suppressed whatever tendencies his DNA might have provided him with that might have made him inclined toward aggressive, violent behavior.

The Neuroscience of Mass Murder

Less than 6 percent of all Boy Scouts ever go on to the loftiest and rarest designation of Eagle Scout. An Eagle Scout has usually spent many years in the Scouts, distinguishing himself by his service in the organization, holding multiple positions of responsibility, and earning at least twenty-one different merit badges in everything from cooking to environment sustainability. Former Eagle Scouts have included Neil Armstrong, Mike Bloomberg, Gerald Ford, and Steven Spielberg. Nine Eagle Scouts have won the Congressional Medal of Honor.

On September 15, 1953, in Lake Worth, Florida, a young Scout was about to make history: he would become the youngest person in the world to achieve the status of Eagle Scout. He was only three months past his twelfth birthday. Thirteen years later, on August 1, 1966, the whole world would know his name: Charles Joseph Whitman.

Whitman was twenty-five years old at the time. He would make his way to the top of the clock tower of the Main Administration Building overlooking the Austin campus of the University of Texas. At 11:45, Whitman opened fire. Having trained as a sniper while serving in the Marine Corps, Whitman had deadly accurate aim. His first shot hit an eighteen-year-old anthropology student, Claire Wilson, in the abdomen. She was eight months pregnant, and the bullet killed her unborn fetus. When Wilson dropped to the ground, her fiancé, Thomas Eckman, ran to her side. Whitman killed him with a single shot. He shot and killed a thirty-three-year-old mathematician. He wounded a graduate student. When a secretary ran to help the student, Whitman shot at her. She sought protection behind the base of a large flagpole made out of concrete. She remained pinned down, out of Whitman's sights, for the next ninety minutes.

Whitman next shot an engineering student in the chest. Then another grad student. Then two women, including another woman who was five months pregnant. He killed two teenagers on a nearby street. Then

he fired at three Peace Corps volunteers, killing one. Then a responding police officer was killed. A doctoral student. An electrician. In the
ninety-plus-minute-long killing spree, Whitman managed to kill seventeen people and wound thirty-two more. Three officers later stormed
the Observation Deck and, in a blazing flurry of pistol and shotgun fire,
finally killed Whitman.

At autopsy, his brain revealed an advanced malignant brain tumor
that had disrupted his prefrontal cortex and undoubtedly interfered with
his ability to restrain his impulses. Earlier that day, Whitman had asked
for medical help and had even begged the police to arrest him before he
killed someone. He was informed he could not be arrested for a crime
that he had not yet committed. The finding of a large prefrontal brain
tumor during the autopsy of Charles Whitman's brain is further evidence that altered frontal lobe function is a component of psychopathic
behavior.

Mass Murder as a Way of Life

Whitman ushered in a new epoch in American cultural history: the age
of the mass shooter. And if mass shootings are an index of the number
of negative zeros in our midst, then that raises some disturbing questions. In the almost fifty years since Whitman shooting spree in 1966 and
the racially motivated killings in Charlestown, South Carolina in 2015,
there has been a fifteen fold increase in the number of mass shootings in
the United States. There seems to a thread that connects many of them.
Consider the following:

On April 16, 2007, twenty-three-year-old Seung-Hui Cho killed
thirty-three people and injured twenty-three more in a shooting spree
at Virginia Tech. His favorite pastime was playing *Counter-Strike*, a first-
person-shooter computer game.

Twenty-four-year-old James Holmes was responsible for the Aurora,
Colorado, cinema shooting on July 20, 2012, killing twelve people and
injuring seventy more. His favorite video game: the violent first-person
combat game *World of Warcraft*, which he played hour after hour.

On July 22, 2011, Anders Behring Breivik, aged thirty-two, killed
seventy-seven individuals and wounded 207 more. His favorite computer

game: the first-person-shooter game *Call of Duty*, which he played for hours at a time, claiming the game had helped him hone his shooting skills.

On September 11, 2003, two brothers, William and Joshua Buchner, ages fifteen and thirteen, respectively, carried out the killing of a random stranger alongside a roadway after being inspired by playing *Grand Theft Auto*, where random roadside killings are a common feature of the game.

An Oakland gang, called the Nut Cases, was responsible for more than 150 killings. Before they went out on a killing spree, their usual modus operandi was for members to get high, then play *Grand Theft Auto* for three or more hours. They would play the game using only the weapons they planned on brandishing for the night's killings. In a single year, the gang killed more than 100 people in the San Francisco Bay Area.

On April 20, 1999, Eric Harris, eighteen, and Dylan Klebold, seventeen, killed twelve classmates and wounded twenty more at Columbine High School. Eric and Dylan were both fanatical fans of the game *Doom*. They were very accomplished players and had hacked into the game and modified it to be able to play together as partners in the same game. They also customized the game rules so they were permitted to carry more weapons, along with an unlimited amount of ammunition. They modified their "opponents" in the game so they would have no weapons and could not fight back. In short, they used *Doom* to practice the killings they planned to carry out at the high school.

There have been epidemiological studies evaluating the connection between the popularity of violent, first-person-shooter games and the national trends in homicide and criminal violence. At the national level, there appears to be little direct correlation. But I am convinced that if one could cull *susceptible* individuals out of the general population, then there would be an association between violent game playing and the expression of aggressive impulses and behavior. First, the litany of mass murderers found to be aficionados of first-person-shooter games is striking. Second, there is abundant evidence that the developing brain will substantially rewire itself in response to computer game playing, allowing the player to become quicker and more adept at the game over time. In addition, brain imaging studies demonstrate that playing violent computer games will dramatically lower the threshold for aggressive, reactive behavior in teenagers.

In one neuroimaging study, a group of typical teenagers was randomly assigned to be in one of two groups. The first group was enlisted to play computer games but was not permitted access to any violent first-person games. The second group of teenagers exclusively played violent first-person-shooter games. Very significant differences quickly emerged on the brain scans of the two groups. First, in less than a single week of game playing, the group exposed exclusively to violent games showed increased amygdala reactivity (see figure 8). In other words, their brains had become primed to exhibit *more* aggressive behavior. Second, they show decreased activity in the prefrontal cortex, the area where the individual looks for self-restraint to control aggressive impulses. I am deeply concerned by the extent and swiftness with which these changes become manifest in the brain. I believe that the reckless marketing of increasingly immersive, aggressive, and violent (particularly first-person point-of-view) computer games enhances the likelihood that young impressionable, developing brains will be inclined to express their increasingly violent impulses.

While it is tempting to play a "blame the brain game," a major societal change appears to be reflected in the number of mass killings since the mass murder carried out by Charles Whitman. It suggests there is widespread and sustained pressure at work in the United States to escalate to mass murder. In fact, the curve may be exponential—the same kind of curve used to describe the spread of COVID in an unvaccinated population. If brains are being changed by the environment, then they are being altered en masse. The latest statistics about homicide among youth are frReference for revised numbers: ightening. While only a minority of our children may become more likely to commit homicide, the majority of them all are at gretare risk of becoming the victims of one as the number of mass shooting and gun-related deaths continues to climb. According to the Department of Justice, one out of every five homicides in the United States is carried out by a minor. Data collected by the Center for Disease Control and Prevention now lists firearms as the leading cause of death among American children. . Gun violence has become a new kind of epidemic that is specifically targeting our children. And now there is a significant, credible link of circumstantial evidence between an increase in aggressive behavior and playing first-person-shooter games.

What is the nature of evil? In 1957, Ayn Rand wrote in her novel *Atlas Shrugged,*

> Thinking is man's only basic virtue, from which all the others proceed. And his basic vice, the source of all his evils, is that nameless act which all of you practice, but struggle never to admit: the act of blanking out, the willful suspension of one's consciousness, the refusal to think— not blindness, but the refusal to see; not ignorance, but the refusal to know. It is the act of unfocusing your mind and inducing an inner fog to escape the responsibility of judgment—on the unstated premise that a thing will not exist if only you refuse to identify it . . . so long as you do not pronounce the verdict "It *is.*"

While the act of feeling and doing evil may not be as volitional as Rand implied, there is neurological truth to her observation that to per-petrate evil, we must enter a state of "inner fog." Brain imaging studies reveal that evil, the ability to wish or do harm to others, arises from a capacity to bypass or dismantle several areas of the brain that help us create relationships and bond with our significant others, including our mates, our offspring, and our friends. As hominids, we evolved potent brain mechanisms to ensure we function well within small social groups, ranging from family to tribal groups. Homicide within these group is considered a crime and an aberration of the covenant between tribal members prohibiting violence.

However, tribal groups are undergoing rapid dissolution. More than half of all the people on the planet currently live in giant cities. This move-ment into our cities has helped magnify the scale of economic disparities, class distinctions, ethnic segregations, pollution, and urban plight. Our brains have simply not had sufficient time to evolve new mechanisms to transport them from the intimate circle of a tribal group to the aston-ishing breadth and pressures of a megalopolis. Nor have these existing mechanisms become sufficiently ingrained in the developing, adolescent brain to make them immune to the onslaught of addictive and immersive violent media that has been unleashed in the digital age. The mechanisms of social restraint that worked for family and tribal groups can also work in larger social units—even those of factions, alliances, political parties, and countries. Perhaps, one day the entire planet. But for now, that sense of being closely bound to one group immediately puts those who dwell outside that circle at risk of being seen as outsiders, foreigners, and even enemies. As we will see, when the brain mechanisms that warm our family and tribal relationships turn exclusionary, they metamorphose into the fires that power the engines of bigotry and hatred.

6

On the Nature of Prejudice, Persecution, and Racism

If only it were all so simple! If only there were evil people somewhere insidiously committing evil deeds, and it were necessary only to separate them from the rest of us and destroy them. But the line dividing good and evil cuts through the heart of every human being. And who is willing to destroy a piece of his own heart?
—Aleksandr Solzhenitsyn, *The Gulag Archipelago*

Gustave Gilbert at Nuremberg

Gustave Mark Gilbert was born in 1911 in New York City, the son of Jewish-Austrian immigrants. He was a bright student and went on to receive a PhD in psychology from Columbia University in 1939. When World War II broke out, he was commissioned as a first lieutenant in the army. Because he spoke German fluently, Gilbert was quickly recruited into military intelligence. In 1945, after the Third Reich collapsed and surrendered unconditionally to the Allies, Gilbert was dispatched to serve as a translator for captured German prisoners as they were brought in to be interrogated and debriefed. Gilbert's facility with the language soon made him a favorite with the German prisoners. Many of the prisoners of war were quick to confide in him, believing his German roots made him far more sympathetic to their situation than other Allied officers sent to question them.

In time, Gilbert's skill at debriefing German prisoners of war got him promoted to the rank of captain. His services were eventually requested to help interrogate several of the most prominent captured Nazi leaders, including Hermann Göring, Rudolf Hess, and Joachim von Ribbentrop. Even in the aftermath of defeat, after he informed them that he was Jewish, nothing seemed to deter these men from sharing many of their innermost thoughts with him.

Captain Gilbert was eventually appointed the chief psychologist to the Nuremberg war crimes trials and given the duties of continuing to deepen his connections with the prisoners and ensure that they did not commit suicide to avoid prosecution.* Thus, at the close of the war, in the wake of confronting some of the greatest mass murderers in history, Gilbert found himself in a unique position, as both a Jew and a psychologist, to try to understand how and why these men could have committed such genocide. In his memoirs, *Nuremberg Diary,* Gilbert wrote, "In my work with the defendants . . . I was searching for the nature of evil and I now think I have come close to defining it. A lack of empathy. It's the one characteristic that connects all the defendants, a genuine incapacity to feel with their fellow men. Evil, I think, is the absence of empathy." Brain imaging studies confirm Captain Gilbert's suggestion that our empathy circuit must be disabled in particular ways when we make up our minds to target specific groups for persecution.

Genocide as a Constant

History is so liberally sprinkled with genocide that it is wishful thinking to believe it is a freak occurrence or momentary aberration. A glance at just the twentieth century is sobering. It opened with the infamous Greek Christian genocide carried out by the Turks of the Ottoman Empire (1914–22); then it was followed by the Armenian genocide (1915–17) and then the Nazi Holocaust from 1933 to 1945. We then move on to the Indonesian genocide (1965–66), then the Pol Pot Cambodian genocide (1975–79), followed by the Kurdish genocide in Iraq (1986–89), then the ethnic cleansing during the Bosnian conflict (1992–95), followed by the Rwandan genocidal killings in 1994, then the Darfur genocide in Sudan (2003—present), then the Yazidi genocide 2014-2019, and ending most

---

* Most notably Hermann Göring was found guilty of war crimes and sentenced to hang. He managed to take cyanide the night before his execution was to be carried out.

recently with the Rohingya genocide (2016—present) and the Syrian genocide being conducted by the government forces under President Bashar al-Assad. This litany should demonstrate that genocide is more like a recurring habit—maybe even an addiction—rather than a fluke.

Dehumanization: The First Step

Many neuroscientists now believe the process of genocidal dehumanization begins with verbal abuse and hate speech. It is almost always the first step taken to differentiate between "us" and "them." Although there is no internationally accepted legal definition of what constitutes hate speech, the United Nations defines it as "any kind of communication in speech, writing or behaviour, that attacks or uses pejorative or discriminatory language with reference to a person or a group on the basis of who they are, in other words, based on their religion, ethnicity, nationality, race, colour, descent, gender or other identity factor." Hate speech serves as an important marker that the process of dehumanization has begun.

Each genocidal episode represents the culmination of successful devolution from dehumanization to extermination. The psychologist Gordon Allport and the philosopher Hannah Arendt have suggested that dehumanization represents a specific process by which one group of individuals willfully strips or deprives another group of individuals of their positive human values. Allport and Arendt went on to postulate that dehumanization is the necessary and sufficient condition that allowed one group of people to feel justified in violating the human rights of targeted individuals, up to and including the use of mass murder. Historically, dehumanization has been a preparatory stage to heighten or accelerate the process by which we lose our empathy for the members of targeted groups and identify them as "others." What is now clear from brain imaging studies is that we don't lose our empathetic responses for those we deem outsiders. We actively bury them.

It is natural for us to feel empathy. Our brains appear to default to compassion. It requires our volition to actively *suppress* our subconscious, empathetic impulses. Neuroimaging studies demonstrate that we "boost" or "restrain" these empathetic responses by the context in which we frame them. For example, we will naturally feel less empathy when looking at a photo of a dour middle-aged man who we're told is a convicted killer than if we are informed that this same man in the photo just lost his wife to cancer. In the former scenario, we suppress activity in our mirror neuron system while we enhance it in the latter. So, the bias established by the *narrative* we are given affects how we will "see" the other person—that is, how much empathy our brains will generate on their behalf. Narrative context contributes to increasing or decreasing our subconscious empathetic responses.

One way that hate speech affects the process of dehumanization— especially now when it can be fueled by inflammatory social media—is by establishing the context that permits us to suspend our empathy toward an out-group. It is natural for us to feel empathy toward our fellow human beings, and that innate, subconscious goodness needs to be actively suppressed to initiate the process of dehumanization.

Brain imaging studies demonstrate there are two separate areas of our medial orbital frontal cortex dedicated to our notions of "dislike" versus outright "dehumanization" of individuals. The brain stratifies the potential "in-groupness" of an individual by looking for differentiating features: "he looks like me," or "he lives in our neighborhood," or "he dresses like me," and so on. We are evaluating characteristics that tell us to which of our own various "in-groups" that individual might belong.

By contrast, if the brain sees little or no commonality with an individual, a different area of the prefrontal cortex will relegate him or her to a dehumanized status. Often, in these studies, photographs of homeless individuals or street addicts will be the ones that trigger powerful reactions of disgust and will be identified by the test subject as "subhuman." Naturally, no one merits relegation to a subhuman status, but brain imaging reveals how our brains segregate the process of disliking someone versus dehumanizing them.

Meta-dehumanization

In chapter 2, we evaluated the "upward spiral of oxytocin." The spiral referred to the mutual reciprocity and entrainment that we see in oxytocin secretion, for example, between spouses in conversation or between a listener and a speaker. Oxytocin creates a kind of "tipping point" that we often detect in our relationships when they deepen and we feel closer and more trusting toward an individual or a group. However, there is also a "downward spiral of oxytocin" to consider too. It arises when we look at how "out-group" prejudice is generated and how the subsequent process of dehumanization occurs.

Research into social psychology reveals that we see groups from which we derive a sense of inclusive identity as in-groups and those with which we have less in common as out-groups. In- and out-groups can be generated along any set of criteria. Race, ethnicity, gender, age, religion, and socioeconomic status are all common characteristics our brains use to define membership in one group or another. But group identity also plays a powerful role in everyday pursuits ranging from sporting events to corporate mergers and even more banal scenarios like gym membership or preferential boarding on airplanes.

The brain is constantly sifting through clues to help us determine if the individuals we encounter are members of our in-group or out-group. In terms of biological evolution, there would be significant survival advantage to being able to develop robust, cooperative alliances in terms of food gathering and mutual defense. And, similarly, it would be helpful if our brains could quickly determine if we were facing an opponent or an ally.

Racial In-Groups and Facial Feature Recognition

We often create strongly held racial in-groups and out-groups. Facial recognition occurs in the area of the brain called the insula. Neuroimaging studies show that the insula is far more attuned to the expressive nuances of our own racial in-group than those of members of other races. Individuals notice that they have a much harder time recognizing the faces of co-workers from different racial backgrounds than they do co-workers from their own race. This issue with facial recognition has profound implications for issues like racial profiling by police. It suggests law enforcement officers would have a diminished ability to interpret the facial reactions of individuals who belong to racial groups other than their own. On a more general note, racial segregation would aggravate issues surrounding facial recognition by reducing the experiential exposure to faces of different races at a critical time in a child's formative development.

Racism is difficult to define. The Nobel laureate Toni Morrison said, "The function, the very serious function of racism, . . . is distraction. . . . It keeps you explaining, over and over again, your reason for being." Such racial discrimination can run the gamut from subtle discomfort to outright hatred toward a race other than our own. Race is one of the most obvious ways the brain can draw distinctions between one group and another.

When we are shown photos of individuals in pain from different races, we generate higher levels of empathetic activity on fMRI (in the insula and anterior cingulate area) if the individual belongs to our own race. Initially, this difference in empathetic activity was postulated to be secondary to sociocultural bias. However, there is evidence that our brains may reserve different oxytocin receptors for those we identify in our own racial group than others. This would mean that for a given level of oxytocin output members of our own race would generate a bigger, more empathetic response. Increased oxytocin activity, as we saw earlier, is associated with heightened trust and empathy.

Subconscious Versus Conscious Racial Bias

Neuroimaging studies reveal we are more racially biased on a subconscious level than we might admit. Researchers obtained fMRI scans from normal subjects while pictures of faces from different racial groups were flashed in front of their eyes. In one arm of the study, the images were flashed so briefly (less than thirty milliseconds) that the participants were not consciously aware of ever seeing the pictures. What researchers found was when non-Black subjects, for example, looked at photos of faces of individuals randomly included in this subconscious modality, there was significantly more activation in the amygdala than when they looked at the faces of Black ndividuals than non-Black ones. For non-Black individuals, this suggests a potentially more volatile, reactive fight-or-flight response being generated subconsciously in response to Black faces. However, these studies reveal an unusual response by which we might overcome these biases too.

The same non-Black subjects were asked to look at the same randomly shuffled set of pictures, but this time the photographic images were allowed to register in the subjects' conscious awareness for a full half second. Now the level of activation of the amygdala was dramatically *diminished*! Instead, there was increased activity in the prefrontal cortex, demonstrating that subconscious threat activation in the amygdala was suppressed when there was adequate time for conscious, cognitive input from the frontal lobes to exert itself. We can overcome innate, subconscious biases by giving ourselves adequate time to cognitively process our responses. Obviously, that cannot happen when our biases are allowed to persist in an unfiltered, subconscious state. Therefore, these imaging studies tell us we need to *think* about racial bias and prejudice rather than allow ourselves to just merely *feel* it.

Tribal Politics and Athletics

How our brains align us with the people around us depends on whether we see them as members of our in-group or cast them as outsiders. For example, when we are asked to evaluate rule infractions in an athletic competition, we are less likely to call fouls on members of our own team (in-group) and far more likely to penalize those committed by our rivals. Another interesting phenomenon is how notions of in-group and out-group extend to the spectators at athletic venues. This self-identification can take on extreme forms. For example, there was a spectator at a boxing tournament who was so upset at seeing his favorite boxer losing a match that he jumped into the ring and came to his boxer's aid by trying to pummel his opponent from behind. This identification with one's team can even turn deadly.

In 1964, a soccer tournament pitted the national Peruvian team against their archrivals from Argentina. Peru was down by a single goal. At the last minute, the Peruvian players rallied and scored a hard-won goal that the referees quickly disallowed. The stadium erupted into chaos. Spectators overran the field. Fights broke out between rival fans. More than three hundred people died in the melee. The police and the army had to be dispatched to quash the ensuing riots. Team sports can get tribal.

Our politics are also becoming increasingly "tribal" and aggressive. But even this may end up being determined by functional traits embedded in our brains. Neuroimaging in young adults has suggested that increased volume in the anterior cingulate area is associated with more liberal politics. One can interpret these findings as suggesting that an emotional awareness may make one lean toward liberal ideologies. At the same time, increased volume in the amygdala appears to be more closely linked to conservative tendencies. Individuals who are more attuned to security issues surrounding threat assessment may be more inclined toward conservative positions. I dread the day when brain scans may tell interested parties which political issues move us or which candidates we will support. However, the dawn of neuro-politics is not far off. While questionnaires and exit polls are notoriously unreliable, brain scans are not.

The Mirror Neuron System: How We See and How We Are Seen

It turns out, however, that issues of bias do not simply have to do with the way we perceive others. Not only do we see their reactions in our own mirror neuron system, but they also reflect back how we believe we are being perceived. As an adolescent, I remember being struck by the famous *Galerie des Glaces*—the great Hall of Mirrors—at the Versailles Palace outside Paris. One of the security guards cracked open one of the mirrored doors and turned it so I could suddenly see an infinite number of mirrors reflected within each other. Our mirror neurons are analogous: we see ourselves in others and others in ourselves. But we also look to see how we see ourselves being perceived by others. Do they think I look powerful or pitiful? Do they see me as innocent or guilty? We react negatively when we believe we are seen as inferior by others. As someone once said, "The best way to get folks to hate you is to let them know you hate them."

If our in-group is perceived as subhuman by an out-group, then it will play a central role in how aggressive we feel toward that out-group. For example, when Americans are told that an Arab or Muslim group has dehumanized them in a survey questionnaire and indicated they felt that Americans were inferior and less culturally evolved than Arabs, then the

Americans are far more likely to respond by dehumanizing Arabs and Muslims in their own evaluation. If, on the other hand, the American test subjects were first told that Arab respondents had revealed that they felt warmly toward Americans and ranked them in very favorable, positive terms in an earlier survey, then American respondents would report favorable attitudes toward Arabs in return.

The same trends in mutual dehumanization have been seen in studies whether it is among Hungarians with respect to the Roma population or among Israeli respondents in reference to Palestinians. It is not that complicated: when we perceive we are considered inferior and subhuman by another group, it virtually guarantees a powerful negative reaction in kind against those who demeaned us. For this reason, dehumanization—particularly if it is detected in hate speech—has to be quickly recognized as a potential trip wire that can unleash escalation. Who is willing to turn the other cheek is always the question.

Putting the Cognitive Brakes on Racial Bias

This research suggests each of us must develop robust ways of detecting dehumanization—whether aimed at ourselves or others. We must learn to avoid our natural tendencies to see ourselves as allied with one camp or another. That doesn't mean we can't have values that make us feel more aligned with one group; it just means that inclusion does not have to come down to a binary choice of simple "in" or "out." Perspective-taking exercises should be regularly encouraged in our educational and social institutions as both developmental and therapeutic tools to undermine any tendency toward dehumanization. For example, how would we feel if our religious identity and prescribed mode of dress were being mocked? Or if our neighborhood was subjected to racial profiling by police? We must recognize an inescapable fact: *the real risk of dehumanization lies in its ability to justify and encourage escalation.*

Let me try to get concrete about why these brain mechanisms matter. In the case of the Black Lives Matter movement, we can begin to see how dangerous it is for a non-Black police officer to be unable to adequately "read" the facial expressions of the Black civilians he may encounter. Does the non-Black police officer misinterpret the fear that

Black individuals feel when they suddenly find themselves at the mercy of the police? Does he misconstrue their profound anxiety as threat? Race unfortunately has dire consequences: about one thousand civilians are shot and killed by police officers each year in the United States. However, if that civilian is a Black male, then he is nearly six times more likely to be killed. A non-Black police officer is five times more likely to discharge their weapon than a Black officer when dispatched to the same Black neighborhood.

These statistics become even more disturbing when we discover that Black individuals who are fatally shot by law enforcement are twice as likely to be *unarmed* as non-Black victims. Furthermore, the militarization of law enforcement and the promulgation of the image of the "warrior police" serve to amplify the sense of threat that non-Black officers may feel. Both of these notions reinforce how the picture in the brain of a non-Black officer might begin to become distorted when seeing Black civilians. They may see Black individuals as being less a part of the citizenry and more as part of an enemy force—one they would normally be inclined to annihilate on the battlefield.

On May 25, 2020, George Floyd, an unarmed Black man immobilized in handcuffs, died in the custody of Minneapolis police. He was murdered by a white police officer named Derek Chauvin. Video of the arrest clearly demonstrated that Chauvin killed his prisoner by kneeling on his neck for more than eight minutes, subsequently cutting off Mr. Floyd's airway until he suffocated. The dehumanizing contempt that Chauvin showed during the prolonged suffering of George Floyd brought about an overwhelming shift in public support for the Black Lives Matter movement. It inspired hundreds of public protests to demand reform of police departments. Why was George Floyd's murder such a pivotal event? Because when millions of other people around the planet saw the gruesome video of his murder, it triggered a profound empathetic identification with Mr. Floyd. He could be seen not as a perpetrator but only as a helpless victim, emblematic of the cruelty of bigotry and violence that the police can visit on Black citizens. That empathetic response was profound enough to make a majority of viewers want to defend George Floyd, to come to his aid, because they identified with him. They suddenly saw him as a member of their own in-group.

Brain imaging studies have repeatedly demonstrated that the more hostile we feel toward a person or group, the less likely our brains will be to cognitively remind us of any sense of mutuality or identification with that individual or cohort. To the extent that discrimination may be the brain's default mechanism for identifying friend or foe, the failure to bring our frontal-lobe-based cognitive reconciliation into play becomes the requisite condition for discrimination to occur. Neuroimaging has also made it clear that the degree to which we are intrinsically motivated to overcome our racial prejudices predicts how much activity our brains will register to correct our own biases. The converse is also true: the brains of subjects who demonstrated the strongest pro-white racial bias will also demonstrate the least ability to detect differences between strangers within those racial out-groups. In other words, the more pro-white racial bias an individual exhibits, the more members of the Black race will—literally—"all look the same" to them. This becomes the double whammy of prejudice because it means that *the more racial bias we exhibit as individuals, the less likely we are to use cognitive processing to overcome it.* But brain imaging has also given us reason to hope.

A unique study demonstrates how potent these frontal mechanisms can be at suppressing our implicit biases. Researchers used fMRI brain scans to evaluate amygdala activation in white participants as they viewed images of black faces. Increased amygdala activity was seen when the subjects were asked to look at unfamiliar black faces but *none* when looking at the photos of popular black celebrities. So, while our amygdala may start firing automatically when we see the image of a random black stranger, there is zero activation if the picture flashed in front of us is of Michael Jordan, Eddie Murphy, or Chris Rock. In short, this strongly suggests that *learned social recognition easily trumps any instinctive impulses our brain might have to otherwise catalog racial in- and out-groups.*

Racial In-Grouping in Infancy

We have now seen sound neuroscientific evidence to challenge the assertion that implicit racial biases are the inevitable by-product of brain wiring. But we can ask, how do such "own race" preferences become wired into our responses in the first place? It would appear to be the predictable result of connections that arise from non-Black subjects to individuals of color. It turns out that the creation of the brain wiring relating to racial in-groups begins in the first months of life. We know, for instance, that babies show marked preferences that match the gender and race of their primary caregiver as early as three months old.

However, when infants are exposed to caregivers of different races, their own race/facial recognition quickly subsides. Other-race exemplars (even those in just picture books) reduced the recognition bias. The "own-race effect" expressed by a baby is really an evolving "expertise" at discriminating the facial features of its caregivers. This is further supported by research with Korean children adopted by other-race families. These children will identify the race of their adopted family as their own racial in-group as adults. If, for example, a Korean baby is adopted by a Caucasian couple, then the child will be much better at differentiating Caucasian features than Asian ones. By the same token, research with African and Caucasian three-month-olds demonstrates that same own-race preferences will occur when infants are raised in their homogeneous in-group environment. However, these same babies will show no preference for racial feature recognition when they are raised in a racially heterogeneous group. Racial preference and facial feature recognition may arise as early as three months old but can be improved or eliminated by raising children in racially heterogeneous groups.

Racial Bias and the Police

Neuroscience has tried to wade into the politically treacherous currents of police violence and brutality against people of color. However, some of the brain-scanning data on racial in- and out-groups might have poisoned the waters. As David Corey, a forensic psychologist and founder of the American Board of Police and Public Safety, summed up the situation, "On the surface, the implicit bias literature is dismally depressing because it tells us that everybody has automatic stereotypes that operate unconsciously and affect behavior." Several important caveats, however, have emerged from brain research.

Jennifer Eberhardt is a MacArthur Foundation "Genius Grant" recipient and a professor of psychology at Stanford who studies how psychological reactions can affect racial relations. Eberhardt carried out a series of experiments in which police officers and university students were evaluated in an fMRI scanner while staring at a dot that held their focus intently on the center of a computer screen. Along the sides of the screen, a random series of images of black and white faces were being quickly and subliminally flashed. Eberhardt then used a well-described technique called a dot-probe paradigm to evaluate the test subjects. The paradigm uses a gradually increasing number and density of dots on the screen until they reveal the discernible image of an object. Subjects are asked to identify the object as soon as they are able to recognize it—remembering that the brain sees what it looks for and not, necessarily, what is there.

In one series of experiments, subjects were "primed" by being shown a subliminal series of Black faces. They were then quicker to recognize a crime-related object (like a gun) in the dot-probe paradigm than a benign, non-crime-related object (like a chair). This association between the images of Black individuals and crime-related objects is bidirectional: if a subject is shown a series of Black faces, they are quicker at detecting crime-related objects; conversely, if they are shown a series of crime-related objects, they are quicker to detect Black faces than non-Black ones.

The data from Eberhardt's study revealed that association between Black individuals and crime is a strong automatic bias in the minds of many police officers. It can also dramatically affect the speed at which an individual officer can "see" a weapon and shoot someone holding one. Eberhardt and her colleagues summed up the findings as demonstrating "that bidirectional associations operate as visual tuning devices by determining the perceptual relevance if stimuli are in the environment." *This finding of an implicit association between black subjects and crime in the minds of white police officers may explain the increased tendency of white officers to imagine they detect a weapon in the hands of a black suspect than a white one.* This would appear to be supported by the disproportionate use of lethal force against unarmed black suspects compared with white suspects.

Justice and Racial Bias

This may explain why non-Black police officers are so often involved in shooting unarmed black men, especially. It is not a justification but an explanation, and to the extent that we can train police officers to understand that Black faces may trigger such associations, it is a step in the right direction. Again, one would hope that targeted training (especially using immersive virtual reality) could quickly and repetitively capture the "priming milieu" of, say, a traffic stop of a Black driver. Police officers could use practice and assessment to evaluate and reinforce those cogni-

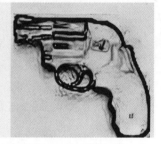

A dot-probe paradigm image of a gun. Subjects were quicker to recognize a pistol if they had been first subliminally primed by being shown black faces rather than white faces.

tive mechanisms that could hold such subconscious impulses in check.

This subliminal priming between Black faces and crime that occurs in the brains of non-Blacks should also give us pause to consider to what extent a Black man (or woman, for that matter) are equal in the eyes of a jury charged with rendering "blind" justice. It is a fact that on average Black criminal defendants receive longer and harsher sentences—including being sentenced more frequently to death—than non-Black defendants. A retrospective review of data derived from adjudicated capital offenses in and around the Philadelphia area showed that the darker the defendant's skin color, the more likely he was to receive a death sentence—an incredibly damning commentary on racial injustice.

Brain Imaging and Better Justice

It is quite clear that the law enforcement and justice systems will need help to elevate their standards of treatment of Black individuals to a higher level before they can gain greater trust in the eyes of the citizenry—for both Blacks and non-Blacks alike. One important contribution that neuroscience is making is using brain scan imaging data to identify situations where police are susceptible to overreacting. For example, the Las Vegas Police Department instituted a policy that barred officers who were actually involved in running a suspect down on foot from handling the alleged perpetrator once the pursuit was over. This produced nearly a 25 percent reduction in the use of force and a 10 percent reduction in injury to officers at the scene.

Why did it work? Because if you are the police officer pursuing a subject, then you will find yourself in a quintessential fight-or-flight situation. Both the person fleeing and the one pursuing will be operating under high levels of amygdala activation. Their physical responses will reflect maximal, exaggerated responses to threat and stress. Under such circumstances of primal amygdala hijacking, the frontal and limbic lobe areas like the anterior cingulate area stand little chance of being "heard"—that is, being able to persuade the individual to restrain his or her behavior. Having a calm, clearheaded fellow officer who has *not* pursued the alleged perpetrator and is *not* under the influence of maximal levels of stress-induced neurochemicals will help ensure that the handling of the subject in custody is carried out in a fashion that is measured and appropriate because *the higher the likelihood of amygdala activation, the greater the potential for overreaction.*

Obviously, the solution to the biased (and often violent) overreaction of police to Black suspects lies in the frontal lobes of the officers. Experienced, highly trained police will "overcome the tendency [to shoot] through the exercise of cognitive control," while inexperienced rookies will not. The more seasoned officers have come to understand that race is a confounding factor and that their concentration in a "shoot/don't shoot" scenario must narrow down on making a quick determination to a high degree of probability that the object in the suspect's hand is, in fact, a gun.

Race is the seismic fault line that runs through the history of the United States. In a 2008 speech to the National Constitution Center, Barack Obama described the U.S. Constitution as "stained by this nation's original sin of slavery." He went on to say that in the hundreds of years that have followed since the founding fathers first drew up the original document, equality could have and should have been perfected and the gap between the real and the ideal closed. But our country has blatantly failed to do so. That history provides me with the single most important implicit racial bias I carry around with me every day as a white man: I do *not* have to think about the color of my skin or the effect it has on others for weeks (or even months) at a time.

Brain imaging does not give us the answer to racial discrimination. That will still prove to be a long and arduous journey for our collective citizenry and its institutions. But neuroimaging does give us the tools to see how racial bias can be a subconscious problem with conscious solutions. It also gives us powerful tools to find out the myriad pathways by which racial bias affects our judgment and decision making. Finally, it gives us ways to analyze how well we bring our empathetic skills to bear on the problems of racial discrimination and injustice. C. Wright Mills, the author of *The Sociological Imagination,* wrote, "Neither the life of an individual nor the history of a society can be understood without understanding both." History is most accessible when it can be personally understood and is only completely understood when all of its constituent voices have the opportunity to be heard. The key to resolving injustice is to genuinely listen to what people have to say about their own lives and personal experiences. That is the message that Daryl Davis delivers in life and George Floyd delivered in death.

7

On the Nature of Sadness, Depression, and Grief

For a petrifying instant here and there, a lightning-quick flash, I want a car to run me over. . . . I hate those feelings, but I know that they have driven me to look deeper at life, to find and cling to reasons for living. I cannot find it in me to regret entirely the course my life has taken. Every day, I choose, sometimes gamely and sometimes against the moment's reason, to be alive. Is that not a rare joy?

—ANDREW SOLOMON, *The Noonday Demon: An Atlas of Depression*

The Epidemiology of Depression

Perhaps there is nothing that defines us more than what makes us feel happy or sad. The *Diagnostic and Statistical Manual of Mental Disorders* (*DSM-IV*)—the handbook of psychology and psychiatry—defines an episode of major depression as a period of two weeks or more during which an individual exhibits five or more of the following symptoms: lack of interest in activities, sad mood, appetite or weight changes, sleep problems, diminished energy, loss of self-worth, guilt, difficulties concentrating, issues with making decisions, and thoughts of suicide or death.

Major depression typically occurs in bouts, and recurrence is the rule rather than the exception. Scientists hypothesize that there may be a "kindling" phenomenon associated with depression, meaning that the first episode of major depression may be triggered by a significant, stressful event in life. But subsequent recurrences can be generated with increasing ease and less specific triggers. Furthermore, the more episodes of depression and the longer their duration, the lower the likelihood of recovery gradually becomes. Because of this kindling theory and the supporting prognostic data, mental health professionals have changed their therapeutic goals from simply reducing the severity of depressive symptoms to helping patients recover back to their pre-depression baselines in the hopes of altering these discouraging rates of relapse. Kindling theory adds a new sense of urgency to developing a coherent public health approach to mental health because the more an individual is allowed to suffer from sustained and frequent bouts of depression without treatment, the more likely the depressive illness will become chronic and resistant to treatment.

This is why the vast under-treatment of depression in America is so worrisome: nearly two-thirds of individuals suffering from depression in the United States do not even seek treatment because they are unaware they are suffering from a treatable disease. Only half of the individuals *diagnosed* with depression get treatment, and only 20 percent receive up-to-date treatment protocols and medications. In other words, if you are suffering from depression in the United States, there is only about a 5 percent chance you will be effectively treated for it.

The vast underdiagnosis and undertreatment of depression in the United States imposes a tragic toll.d The numbers are staggering: twenty-five million Americans suffer from depression. In 2014, there were nearly forty-three thousand suicides reported; more than half occurred in individuals suffering from major depression. David Foster Wallace, the author of *Infinite Jest,* suffered from depression. In 1997, he was awarded a MacArthur Fellowship and wrote quite movingly about depression and its strong connection to suicide:

> The so-called "psychotically depressed" person who tries to kill herself doesn't do so out of quote "hopelessness" or any abstract conviction that life's assets and debits do not square. And surely not because death seems suddenly appealing. The person in whom [this] invisible agony reaches a certain unendurable level will kill herself the same way a trapped person will eventually jump from the window of a burning high-rise.

Wallace himself committed suicide in 2008.

Serotonin and Depression

For more than fifty years, it has been postulated that a chemical imbalance is the underlying cause of depression. It has been hypothesized that low circulating levels of the neurotransmitter *serotonin* are the cause of that imbalance in the brain. This theory has been based on several findings. The first was that medications like monoamine oxidase inhibitors, and then, later, a newer group of drugs called selective serotonin reuptake inhibitors, made depressed individuals feel better. Despite the diverse mechanisms by which these various classes of antidepressant drugs worked, all of them seemed to make more serotonin available within the synapses of the brain. And while that might be true, we cannot directly measure the exact concentrations of neurotransmitters in the brain. Most of the "proof" for serotonin's role in depression is based on the second-hand evidence that the best pharmacological responses are seen with drugs that increase the availability of serotonin. The leading hypothesis about the neurochemical imbalance underlying depression is that there are sufficient concentrations of serotonin in the synaptic clefts of the brain.

The best agents on the market today for the treatment of depression are these selective serotonin reuptake inhibitors, or the newer version, called norepinephrine and serotonin reuptake inhibitors. These agents all interfere with the reuptake mechanisms that normally terminate serotonin's chemical interactions in the synapse. These uptake inhibitors work well and effectively for most patients with depression—but not all, as we will soon discuss.

In medicine, it is widely acknowledged that surgeons can be arrogant—I admit it. And we have also been described as being "seldom wrong but never in doubt." But the same could be said about the psychiatrists who, for more than a generation, have forged this connection between low serotonin and depression. But their case is largely a circumstantial one. As one scientist observed,

The serotonin hypothesis of depression has not been clearly substanti-

ated. Indeed, dogged by unreliable clinical biochemical findings and
the difficulty of relating changes in serotonin activity to mood state,
the serotonin hypothesis eventually achieved "conspiracy theory" status, whose avowed purpose was to enable industry to market selective
serotonin reuptake inhibitors to a gullible public.

While there is data that suggests that depression may be associated
with low serotonin activity, that does not mean it is the cause. We have
no chemical test available to diagnose depression. But these days, some
of the most exciting data on depression is emerging from brain imaging
research. One study evaluated more than a hundred individuals suffering
from untreated depression. They all underwent baseline brain scans and
were then randomized to be treated with either a frontline antidepressant or traditional psychotherapy (but without medications being given
during the study). After twelve weeks of treatment, half of the subjects
reported they no longer felt depressed. The other 50 percent of subjects
who failed to respond to one form of treatment were then allowed to
cross over to the other treatment arm. Despite this, 30 percent of the
test subjects failed to improve with *any* therapy. However, this group
showed a distinct pattern of increased activity in an area of the cingulate
lobe called the subcallosal cingulate cortex. This may be a critical finding because it suggests that we are still missing some vital information
in the story of depression. Brain imaging may help evaluate what kind of
therapy might work best for an individual patient. There is also another
tantalizing clue.

Electroconvulsive therapy—often called shock therapy—has been
around in medicine for more than half a century. We know it induces a
kind of generalized seizure in the brain but still have no idea exactly why
it works. However, electroconvulsive therapy has proven to be very helpful in salvaging some patients with *intractable depression,* that is, resistant
to treatment. Now neuroimaging studies may reveal why electroconvulsive shock therapy works in such cases: it appears to dramatically reduce
activity in the subcallosal cingulate cortex (SCC). In a recent brain fMRI
study of more than a thousand individuals, artificial intelligence analysis
revealed that heightened activity in the SCC was one feature that seemed
to characterize those patients who failed to respond to traditional anti-

depressant therapy or medication.

The study looked at differences in brain connectivity among a cohort of depressed individuals. The scans of a large number of normal, non-depressed subjects were included as controls. Artificial intelligence was used to sift through millions of images looking for any detectable differences. The artificial intelligence algorithms mapped the volumes of connecting fiber bundles on all the scans. This is the kind of task that no human could ever do, but it is precisely the kind of so-called deep learning that computers excel at executing. That is because today's powerful computers can carry out image processing on hundreds of thousands of scans and make hundreds of minute digital measurements on each one—all in a matter of milliseconds. So, it was not human observers who made the next discovery. It was artificial intelligence that did.

The computers learned to identify four different subtypes of depression, based solely on the fMRI maps they were reading. The imaging analysis carried out by the artificial intelligence algorithm found two imaging patterns that were typical of patients suffering from so-called anhedonia. These were individuals whose major depressive symptoms were lethargy and diminished engagement; they derived little or no pleasure from life—unfortunately, a common symptom in depression. But the computers were not done.

The deep learning image analysis went on to discover two more patterns characteristic of the kind of depression that is marked by a preponderance of *stress- and anxiety-related* symptoms. These patients often exhibited other related symptoms like pessimism and restlessness as prominent features of their depression. The artificial intelligence algorithms not only became proficient at evaluating fMRI scans but could accurately predict if a subject was in the cohort of depressed patients or in the normal control group. AI was more than 90 percent accurate at making the correct diagnosis. Without deep learning by computers, we would never know these new imaging patterns existed. AI and neuroimaging are now giving us new paradigms for diagnosing depression and also assessing the brain's responsiveness to a variety of treatments. The time is not far away when we will be diagnosing people with mental health problems with brain scans and evaluating their responses to therapeutic intervention through AI. The application of AI to fMRI studies has helped to identify unique patterns of brain activity associated with subtypes of depression.

What do we have when all this new information is taken into consideration? We are still left with just a *theory* about depression. That's all. No matter how dogmatic and widely accepted that theory may be, a direct link between serotonin levels in the brain and depression is still lacking. The hypothesis *does* help us to explain a lot of pharmacological responses to antidepressants. However, we are left with some unsettling findings. The first is that anywhere from one-third to half the patients with depression are nonresponders, meaning that what we have to offer doesn't work well or at all. That's a very sizable number of people: given current estimates that approximately twenty million Americans suffer with depression, that could be anywhere from six to ten million people who are unlikely to get adequate relief from their depressive symptoms. In addition, as we mentioned earlier, the more bouts of depression one has, the more recalcitrant the disorder becomes. The situation may get worse because, lacking a viable hypothesis about the underlying target for pharmacological agents, many pharmaceutical companies have abandoned basic research. But the imaging is telling us a lot more about the nature of depression.

Structural and Functional Changes in the Brain with Depression

Significant memory lapses are a prominent feature of the clinical picture of many individuals suffering from chronic depression. The reason why has only recently become clear with functional brain scans: individuals who suffer from chronic depression have significantly smaller hippocampi than age-matched controls. While this loss of tissue in the brain's memory centers helps explain the lapses, scientists must now wrestle with why shrinkage occurs in such a vital area of the brain in the first place.

A leading suspect in this emerging puzzle of depression may be stress. For decades, it has been known that rich connections exist between the hippocampus and what is called the hypothalamus-pituitary-adrenal (or HPA) axis. The two areas are immediately adjacent to each other. The HPA axis is the body's general alarm system and controls the outflow of stress-related hormones, like cortisol. Individuals suffering from depression also suffer from chronically high levels of stress. They exhibit sustained elevations in serum cortisol, and eventually they will exhibit significant dysregulation of the HPA axis.

One of the earliest laboratory tests that doctors used to diagnose depression evaluated a patient's HPA axis. Often the adrenal glands of depressed patients were operating at such full capacity for such extended periods of time that it would lead the HPA axis to show evidence of abnormal regulation: it would simply no longer shut down. The HPA in such patients was operating like a runaway train.

The second clue came from examining cortisol receptors in the hippocampus of such HPA-impaired depressed patients. In normal subjects, the hippocampus exhibits an exuberance of neural connections with the hypothalamus. Scientists discovered that the more cortisol was bound to receptors in the hippocampus, the fewer connections would exist between the memory center and the HPA.

One current hypothesis is that exposure to chronically elevated cortisol inhibits the ability of the hippocampus to develop rich intercon-

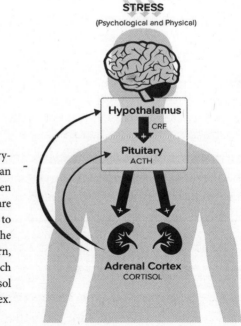

The hypothalamic-pituitary-adrenal axis. There is an intimate connection between how outside, stressful events are processed within the brain to bring about signal changes in the hypothalamus, which, in turn, controls the pituitary, which modulates the outflow of cortisol from the adrenal cortex.

nectivity. This could explain the structural changes in the brain and the memory lapses seen in depression. Corroborating these imaging findings are autopsy studies that reveal that the brains of chronically depressed individuals demonstrate much smaller, denser hippocampi that appear to have suffered damage from elevated inflammatory chemicals (called cytokines) that circulate in the bloodstream but can impair or actually destroy brain cells. Chronic depression can be associated with sustained, unrelenting stress and elevation of inflammatory markers that may, in turn, bring about long-lasting damage to brain structures.

The puzzle of depression is far from solved, but some pieces are beginning to assemble in a tantalizing fashion. First, we have the evidence of a shrunken hippocampus, linked to an overstimulated hypothalamus that is flooding the body with stress hormones. These lead to elevated inflammatory markers that eventually attack the brain and its functional connectivity. So, we can ask, what happens when we treat depression with a traditional serotonin reuptake inhibitor that increases the amount of serotonin circulating in the brain?

We know many patients with depressive symptoms will improve dramatically on such medication. What happens to their hippocampi when we treat them with medication? The answer: autopsy studies of the brains of patients with depression who were treated with serotonin reuptake inhibitors show a restoration of hippocampal volumes and a return of a brain blessed with healthy dendritic connectivity! Therefore, these patients represent true successes at restoring normal brain function and emphasize why it is so critical that individuals with depression have ready access to pharmacological therapy at an early stage in the development of the illness. Early intervention is the best way to avoid the later systemic and anatomic changes in the brain that are becoming the new hallmarks of depression. However, an entirely new and unexpected challenge about the use of antidepressants is beginning to worry clinicians.

Antidepressants and the New Threat of Addiction

Unfortunately, there is now a new and very real threat from serotonin reuptake inhibitors: addiction. Until very recently, antidepressants were marketed and prescribed with the notion that they *never* produced addiction. However, recent findings suggest that is far from the truth. Serotonin reuptake inhibitors appear to have the potential of inducing withdrawal. Sadly, many of the individuals most at risk may be in our own families. Why is this creating such a stir? Imagine what the following statistics may mean in terms of the magnitude of the serotonin reuptake inhibitor crisis. The global market for antidepressant medications is worth nearly $17 billion. Recent research reported by Harvard Health Publishing indicates that individuals who take serotonin reuptake inhibitors are most susceptible to developing withdrawal symptoms when trying to discontinue their medications.

The chronic use of serotonin reuptake inhibitors will cause the brain to downregulate the amount of serotonin it produces. But this also means we are now leaving the brain vulnerable to withdrawal symptoms if the serotonin reuptake inhibitors are abruptly halted or withdrawn. The pharmaceutical industry has not been eager to see the golden goose of chronically prescribed serotonin reuptake inhibitors crippled by concerns over addiction. It has little stomach for seeing a thorough investigation done of the increasing number of individuals suffering from withdrawal symptoms after discontinuing antidepressants. It's simply not good for marketing, and it discourages future consumers.

According to a 2018 article in *The New York Times,* nearly sixteen million Americans have been on antidepressant medications for more than five years. This rate of chronic, sustained use of antidepressanats has doubled since 2010 and tripled since 2000.. Only some individuals on serotonin reuptake inhibitors can simply quit taking antidepressant medications without experiencing withdrawal or what psychiatrists are currently labeling "discontinuation syndrome"—medical-speak for addiction withdrawal. This kind of language is akin to telling someone who is being fired that they have been placed on "an indefinitely extended hiatus." The chronic use of antidepressants may be causing addiction, and the abrupt cessation of serotonin reuptake inhibitors can be associated with symptoms of withdrawal.

All of this is taking place against the backdrop of physicians having prescribed these drugs for years to people who were assured these medications were safe and nonaddictive. In fact, most pharmaceutical studies seeking approval for new antidepressant medication from the Food and

Drug Administration (FDA) typically evaluated the effects of using the medications for little more than a month or two at most. It was never anticipated that the use of these medications over years might require a whole new reassessment of the dangers they represent. Until recently, government regulators from the FDA were not even collecting data on reports of withdrawal issues in patients taking antidepressants.

Psychiatrists, however, are increasingly cautious that patients may need to be gradually weaned off their serotonin reuptake inhibitors over an extended period and may even require the use of other medications to help them through the issues of withdrawal. I personally know of five patients who have been unable to discontinue antidepressants because of the severity of withdrawal symptoms they experienced. This is not a minor problem.

For example, in New Zealand, 75 percent of chronic antidepressant users reported feeling withdrawal symptoms when they tried to stop their medications. This creates an enormous public health problem given that, back in 2016, there were thirty-five million Americans taking antidepressants, according to the National Health and Nutrition Examination Survey. There are also significant gender issues to consider because one out of every five women in the United States currently takes an antidepressant—a rate more than double that of their male counterparts. In addition, antidepressant withdrawal may now be creating a substantial impediment to many patients being able to be taken off their medications.

Grief and Bereavement

Grief represents one of the saddest experiences any of us will endure. And it's universal. While not all of us will have to struggle with depression, every single one of us will come to know grief. And we will come to see that it is not just sadness. As C. S. Lewis wrote, "No one ever told me that grief felt so like fear." And contrary to popular myth, you never "get over" grief. It creates a hole—a yearning—in your life that never fully resolves. Grief is also isolating and lonely. Dr. Katherine Shear, a psychiatrist at Columbia University, writes that "in spite of the shared experience and strong social support, most bereaved people feel more alone than at any time in their lives."

Brain imaging makes it clear why grief is such a tumultuous experience that leaves us depleted and confused. The British psychologist John Bowlby points out that what makes grief different from depression is that "during this [grieving] process our minds naturally, and mercifully, oscillate between confronting and avoiding . . . the painful reality." As we will see, what Bowlby describes correlates with what we can visualize in the functional imaging of the grieving brain.

As a rule, clinicians typically divide grief into two categories: uncomplicated and complicated. I don't quite get the distinction myself, because, as you will see, all grief is complicated from the brain's perspective. However, from a therapeutic perspective, uncomplicated grief supposedly leaves you overwhelmed with sadness and loss, but—after some arbitrarily defined interval—you appear to be outwardly able to pick up the pieces of your life and return to some semblance of your usual functional capacity. While in complicated grief you don't do as well. You fall apart and the grief spirals into becoming such a central tenet of your life that it leaves you believing that there's no hope life will ever be any good anymore. Rhonda O'Neill, who is a nurse and the author of *The Other Side of Complicated Grief,* put it very eloquently:

> Grief is a game changer in life. Not only have we lost someone who was a vital part of our lives, we have lost the future that we so naively believed was going to unfold before us, just as planned. We are left with an uncertain, and ill-defined future, and no instructions as to how to get to a place where there seems to be any future worth living.

To say grief is stressful is an understatement. It is one of the most singularly stressful events in a person's life. And dangerous. Losing a spouse or a child is associated with a mortality rate of 3 to 8 percent within the first year after the death of the loved one. There is now substantial documentation of the so-called broken-heart phenomenon. This is a stress-induced cardiomyopathy that induces heart failure. A dramatic increase in stress-related hormones is seen quite frequently among grieving survivors and produces an effect like a heart attack. Grief represents one of the most significant stresses in life and can carry with it a significant increase in morbidity and mortality.

Neuroimaging scans now clearly demonstrate that grief is also characterized by an oscillatory state where the brain activity alternates between its reward system and its pain perception system. For example, when grieving subjects are placed in an fMRI scanner while photos of the deceased are flashed in front of their eyes amid a randomly shuffled deck of unrelated images, the grieving brain shows two different results. Sometimes, a photo will induce increased activity in the pain network. Then, minutes later, when the very same images come around again, it can instead produce heightened activity in the nucleus accumbens, the pleasure system. So, the same picture of the deceased loved one will now produce sensations of reward and longing. The changes in brain activity are strikingly similar to the ones we see when people fall in love. That should not shock us. When we fall in love, we experience an intense longing to be with the object of our love that can also oscillate between pain and pleasure. And grief is no different. Brain imaging analysis of grieving individuals corroborates alternating activity between the pleasure and the pain centers of the brain.

There are other oscillations between different centers of the brain that help distinguish grief as a kind of "binary" state. In recently bereaved subjects, brain scan studies reveal high activation levels in the amygdala that seem linked to the degree of sadness the subjects report feeling during a particular imaging session. Remember: the amygdala serves as one of the brain's epicenters for stress. The intrusiveness of grief reported by bereaved individuals appears to correlate with *increased* activity seen in the amygdala and the anterior cingulate area. The ability of individual test subjects to avoid dwelling on their grief—to be able to push it aside temporarily—correlates with how well the subjects' prefrontal cortex can suppress activity in the amygdala. This matches well with other brain scan studies that demonstrate that the anterior cingulate area serves as the emotional gas pedal. At the same time, the prefrontal cortex acts like cognitive brakes to keep emotions in check. This is the same combination we saw in the previous chapter on prejudice where the prefrontal cortex helped the individual suppress subconscious impulses of racial bias.

Few things in life can be more tragic than the loss of a child. Neuroscientists have begun to use brain imaging to evaluate this particularly devastating state of grief. One study evaluated twelve women who had lost an unborn child and compared their brain scan results with twelve women who had delivered healthy babies. The photographic stimuli used were generic pictures of happy babies that were intermingled with the faces of adults with happy or neutral expressions. I can only imagine how hard such a study must have been for the bereft group of mothers to undergo.

The brain scans of the mothers who had lost a baby demonstrated enhanced activity when looking at the pictures of babies in a unique pathway between the thalamus, where we typically process physical pain, and the cingulate area, where emotional pain is managed. This suggests that in the brains of these bereaved mothers grief took on both a physical and a psychic dimension. Researchers found, contrary to their expectations, that the grief experienced by mothers who had lost their babies five to seven years prior to undergoing brain scanning was as intense as that of mothers who had just lost their child within two weeks prior to being imaged. This only reinforces the notion that while people may adjust to grief, that does not mean the trauma and pain of that loss are necessarily attenuated with time. Brain imaging studies now suggest that grief may remain undiminished for years after the loss of a loved one.

Another differentiator between grief and depression comes down to the neurotransmitters involved. As we just discussed, the basic underlying neurochemical mechanism in depression revolves around the paucity of serotonin available to the brain. Grief involves far more neurochemical markers, including dopamine, endorphins, and oxytocin, and often represents a state where these neurochemicals are lacking or at least diminished.

With the loss of a loved one, the withdrawal of these neurotransmitters translates into a state of enhanced sensitivity to pain and discomfort and a loss of well-being and trust. Acute grief can therefore lead us to feel overwhelmed and threatened. It can thus trigger a fight-or-flight response with sustained activation of the whole HPA-related cascade of stress hormones along with the abundant inflammatory biomarkers and depressed immune function. The difference between uncomplicated and complicated grief would seem to come down to this: How long is this stress response sustained? When the stress response and subsequent immunosuppression become chronic, grief can easily devolve into a more complicated emotional and physiological form that can include depression as a component.

As of this writing, it is estimated that as many as 1.2 million Americans have died from coronavirus in the last four years. Let me try to put the dimensions of this loss—this national grief—into context: that is more than all the soldiers, sailors, airmen, and marines America lost in World Wars I and II combined! As a scientist, this massive loss of life is all the more bitter and cruel since 80 percent or more of those losses were entirely avoidable had federal leadership been strong enough and inspired enough to marshal the will of the American people early in the evolution of the pandemic. Every one of those tragic deaths will leave a number of individuals behind to suffer grief and loss. In the age of COVID-19, this global sadness is deepened by our inability to be at the bedside of our loved ones when they pass. There is no opportunity to say the final goodbyes. No chance to rally the loved ones at the deathbed. No funeral service. No eulogies. No wake. No breaking bread. No shoulders to cry on. No songs. No shared stories.

Early on in the course of the pandemic, I watched a news story on television about a priest in Italy making his way among almost a hundred coffins of COVID-19 victims. Each had died alone. In an ICU. On a ventilator. Each one left behind a heartbroken family. You *know* there were tears being shed in distant rooms, around dinner tables with an empty chair. Clothes still hung in the closet, holding the loved one's scent in their weave. The book still laid open on the bedside table. The sound of their recorded voices still taking messages on their cell phones, vowing to call you back as soon as possible. I know. My mother died during the COVID pandemic. My brother could not come to see her. He had to say his goodbyes via FaceTime while I held up my cell phone. It was as if her life weren't over. It was merely interrupted. A bad signal. An unfinished phone call that would be resumed when the reception was better.

The priest I was watching only paused long enough at the foot of every coffin to sprinkle holy water on it. It was the only hint of any ceremony whatsoever. COVID-19 reminds us that there are more painful ways of dying. Circumstance can impose upon us death without closure. Loss without ceremony. It's death in a vacuum. It's like death happening to someone trapped in the soundless solitude of a space capsule.

8

On the Nature of Happiness

Now and then it's good to pause in our pursuit of happiness
and just be happy.
—GUILLAUME APOLLINAIRE

Happiness: What's Missing?

I once took care of a fisherman from Gloucester, Massachusetts. He was twenty-eight years old and worked as part of the crew on his father's fishing boat. The boat had been dragging a large fishing net behind it when the net suddenly became snagged on the ocean bottom. The men tried to pull it free and dislodge it. Finally, in desperation, they hooked up one of the winches on the end of a large ten-ton metal boom and started trying to hoist the net up using the additional horsepower of the power winch. It started whining away and smoke started to rise from the housing. Then, suddenly, the metal cable snapped under the load. The boom swung around explosively and hit the fisherman at full speed, straight in his head.

The stricken man came into the emergency room at the Massachusetts General Hospital with the top part of his skull shaved off and his brain exposed. I remember how perfectly astounding it was to see this man completely alert, sitting up, talking to me as I watched his brain pulsate with each heartbeat in his skull. He started coughing and politely put his hand up to cover his mouth. As he did, a few bits of his brain sloshed over the ragged edge of the exposed skull. It then ran down the side of his head like wax dripping from a candle. I covered the open wound with a sterile towel soaked in antibiotic solution. "If you can, could you please try to stop coughing?" I asked him. I don't think he understood why.

We took the man to the operating room. We labored for hours and were finally able to stop the bleeding and, with our plastic surgeon's help, close the scalp. He survived the surgery. Unfortunately, he lost a great deal of his frontal lobes. Several months later, his father brought him back to the clinic for a follow-up visit with me. He could no longer be trusted to drive himself.

When I asked his father how his son was doing, the father answered, "Fine. He likes to go out on the boat now. Mostly he sits out on deck up in the bow. He seems to like the sun and the wind and the sea." Then he slapped his son on the shoulder. "You didn't always like that, did you?"

His son shook his head and smiled.

The father explained, "He used to hate going out on the boat. All he dreamed about was leaving . . . getting out. He used to tell me he was going to leave Gloucester. Pack his car, hit the road, and never look back." He patted his son and looked at me: "Well, at least he's a happy fisherman now."

That comment sent a chill through me. Was he really happy now, or did he simply have no choice? What part of his brain was lost? Maybe the part that could choose freedom. Some cases haunt you your whole life as a surgeon. This one does. Because, yes, I did what I was supposed to do: I saved his life. But it was turned into a life without the capacity to choose. Can a life ever be happy without the freedom to choose?

The Neuroanatomy of Happiness

Neuroimaging has now revealed how the responses of pleasure and happiness in the brain play a significant role in how we learn to make choices that make us happy. Take an experience like playing basketball. For example, let's say you are recognized as an outstanding basketball player in your high school. The experience of that acknowledgment activates the same hedonic substrate of dopaminergic pathways we looked at in the earlier chapters on lust and love. The *pleasure* you associate with being acknowledged by your high school peers in a basketball game becomes tightly linked in the brain with a "like" hot spot that links dopamine release to basketball playing. This linkage, in turn, will motivate you to play more basketball because it is an activity that you have *learned* makes you feel *happy.*

While the pleasure system is a diffuse, far-flung network of adrenergic and dopaminergic fibers, the pleasure-causation system is a much smaller one with only a handful of these "like" hot spots in it. To date, these distinct "like" hot spots have been identified scattered about in the brain: in the outer layer of the nucleus accumbens, ventral striatum, and a small area in the brain stem. These hot spots are organized in a cooperative hierarchical fashion so that multiple hot spots can fire in sequence and then join forces. The more "like" hot spots fire together, the stronger the individual's "like" reaction will become and the greater his or her motivation will be to pursue that activity.

This mechanism is vital to understanding why one person may like math class and another loves horseback riding or astronomy. And while it has been said "there is no accounting for taste," we can at least explain how the brain acquired that taste through these "like" hot spots. It also has profound implications about school, in particular. We can see that helping a young person develop into a successful student involves searching through a wide range of activities that can engender (and, later, entrain) "like" responses. It comes down less to encouragement (although that's also important) and more to the gratification derived from activities that engender personal success. "Like" hot spots in the brain work in hierarchical fashion to ensure we learn what activities bring us self-satisfaction and happiness.

This creates an important part of the distinction the brain makes between a hedonic response and happiness. For example, individuals who are addicted to heroin may feel pleasure when they get an intravenous dose of the drug, but they don't feel happy about being addicted. By the same token, I might get pleasure from playing a game of catch with my son. Each time one of us sinks a ball into the other's baseball glove with speed and accuracy, there's pleasure registering in our brains. But there's also happiness that is derived from the two of us—father and son—being joined in playing catch together in the driveway, as we have done for decades since he was a little boy. The *pleasure* of catching a ball helped teach us about the *happiness* of playing ball.

Understanding of the neurophysiology underlying the hedonic response is still in its infancy, but already imaging studies reveal dramatic insights into the role that specific areas of the brain play in how we interpret our experiences. The very inside, or medial edge (facing toward the center of the brain), of the mesio-orbital frontal lobe (MOFL) seems to encode for positive responses to emotional events like a graduation (see point D in figure). The lateral region of the MOFL is activated by unpleasant or disgusting responses where the individual may want to escape a claustrophobic environment, for example (see point C), or a humiliating event. This side-to-side (medial-to-lateral) gradient varying from pleasant to unpleasant in the MOFL is cross-registered with a front-to-back (anterior-posterior) gradient from abstract to concrete, respectively. So, a symbolic reward—like a wad of money (see point A)— would produce a more anterior response in the MOFL. A more physical reward—like a good-tasting slice of cheesecake (see point B)—would be a more concrete sensory appeal and therefore produce activity in the more posterior aspect of the MOFL.

These subtleties of brain function—of how the brain gets into the granular detail of assessing and cataloging experience—are just coming

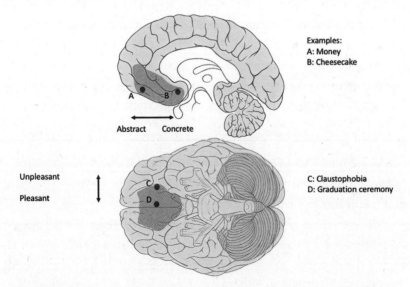

A schematic of a sagittal view of the brain in the upper part of the panel and a view of the underside of the brain in the lower part of the panel. The MOFL is the dark shaded area of the brain to the left.

to light with neuroimaging.

Brain imaging studies also show that happiness and sadness are relegated to different sides of the brain. Functional MRIs showed higher
activity in the *left* MOFL when people were reporting positive, happy
feelings, while the MOFL on the *right* side showed higher levels of activity while experiencing negative or sad feelings. There appears to be significant lateralization and regionalization in the frontal lobes when it
comes to happiness.

The Happiness Set Point

Most of us believe that our situation, good or bad, determines whether
we are happy and that happiness is relative. In the late 1970s, the psychologists Philip Brickman and Dan Coates from Northwestern University
and Ronnie Janoff-Bulman from the University of Massachusetts asked
a seemingly obvious question: "Who do you think would be happier:
a person who had just won the lottery or someone who had recently
become paralyzed in an accident?" A no-brainer, right? Brickman and
his colleagues evaluated twenty-two major lottery winners, twenty-two
controls, and twenty-nine paralyzed accident victims. The victims had
all become paralyzed within the preceding year and included eighteen
quadriplegics and eleven paraplegics. The lottery winners had hit the
jackpot within the last two years.

Interviewers asked the research subjects three questions: How happy
were you in the past? How happy are you, right now, at this stage of
your life? How happy do you expect to be six months from now? At first
glance, it would seem easy to predict how the answers would go. Lottery
winners should feel great. Paralyzed victims should feel terrible, and the
control group should feel, well, all right. Just average but okay. However,
adaptation theory predicts a different outcome.

The American psychologist Harry Nelson originally developed
adaptation-level theory in the first half of the twentieth century to
explain how humans adapted to light after being in a dark room. It simply states that "people's judgments of current levels of [light] stimulation
depend on whether this stimulation exceeds or falls short of the level of
stimulation" to which they have become accustomed. If we translate this

theory from a dark room to real life, adaptation theory would predict that individuals who have a stroke of good fortune, like winners of a lottery, would be no happier than any other folks. There are two reasons that adaptation theory makes this prediction. First, after winning something like a multimillion-dollar lottery—a relatively huge "high" in an average person's life—almost routine "happy" events in life would be a letdown. As one of the authors of the study wrote, "While winning $1 million can make new pleasures available, it may also make old pleasures less enjoyable."

Naturally, those individuals who won the lottery rated it as a very positive occurrence in their lives, while paralyzed patients rated their accidents as very negative events in theirs. However, there was no difference between winners and controls in how happy they rated themselves six months earlier, now, or how happy they expected to be in the future! Accident victims rated themselves as much happier in the past than controls, and rated themselves as less happy in the present than controls. *But* the paralyzed subjects rated themselves equally optimistic about how happy they thought they would be in the future. In retrospect, accident victims understandably saw themselves as being happier in the past, before they were paralyzed compared with their present situation. However, when it came to life's everyday pleasures and rating how optimistic they were about the potential they saw for being happy in the future, it did not matter if one had won the lottery or been paralyzed. The lottery winners were no happier in the past nor in the future than individuals who had become paralyzed in an accident.

The implications of this study are wide-ranging. For example, we can predict that individuals raised in underprivileged circumstances should be inherently no less or more happy than individuals raised in a privileged setting. Similarly, individuals suffering with blindness or disabilities would be predicted to be no less happy than able-bodied individuals. And despite the nostalgia that people have for how happy they were in the "good old days," studies reveal that people in the 1940s were no happier or unhappier than people in the 1970s. Variables such as age, gender, socioeconomic status, and education have little effect on how people rate their happiness.

We might assume that much of happiness can be derived by comparison. Expressions such as "the grass is always greener on the other side of the fence" would suggest that happiness implies a relative or comparative value in the eyes of the beholder. One study by Daniel Kahneman of Princeton University asked a group of students from the Midwest, and another from Southern California, to rank themselves with respect to a happiness inventory. They were also asked to decide which group of students were more likely to rank themselves as happier and with more opportunities. Both groups opined that students in Southern California— you know, suntans, beaches, and surf versus cheese, snowblowers, and opening day of deer season—would be the happiest. In fact, individual happiness inventories revealed the two groups of students were equally happy. It is still unclear why some individuals seem to be innately uplifted by the glass half-empty and others grow despondent by the same glass seen as half-full. That answer may lie in our DNA itself.

Genetics and Set Points

David Lykken (1928–2006) was a controversial psychologist at the University of Minnesota, where he headed up the Minnesota Center for Twin and Family Research (MCTFR). He evaluated the heritability of certain psychological traits by comparing fraternal twins (who are born together but *do not* have identical genes) with identical twins (who share the same genetics). Lykken studied more than twenty-three hundred members within the MCTFR twin registry. He evaluated the twins with respect to education, socioeconomic status, and rankings for contentment and well-being. They were also retested ten years later. In addition, by evaluating the scores of one twin, Lykken was able to accurately predict what the other twin would report as scores and how they would score themselves a decade later.

Lykken put forward a theory that individuals were born with a *set point of happiness*. It was sort of like a thermostat setting of how happy a person would feel. From his twin studies, he calculated that at least 50 percent of wherever that happiness set point might fall was predetermined by genetic makeup. Even though we think of them as large contributors to our personal sense of happiness, factors such as socioeconomic status, level of education, income, marital status, even the degree of religious devotion accounted for less than 3 percent of the variance observed in the reports of well-being submitted by the twins. So, the major portion of our happiness and contentment is determined by heredity. Lykken's sobering conclusion was this:

> If the transitory variations of well-being are largely due to fortune's favors, whereas the midpoint of these variations is determined by the great genetic lottery that occurs at conception, then we are led to conclude that individual differences in human happiness—how one feels at the moment and also how happy one feels on average over time—are primarily a matter of chance.

This offers us additional insight into why winning the lottery or being paralyzed—as major life events as they represent—still do not overwhelm our genetically determined set point of happiness. We are like corks bobbing upon a sea of circumstance. We may rise or fall on the waves of fortune, but sooner or later we will return to floating at the same level where we started. More than half of our set point for happiness appears to be genetically predetermined.

Happiness Is a Good Wine

We are constantly the victims of our choices. A wine label will help prove my point. In 2007, Brian Wansink, a researcher from the Food and Brand Lab at Cornell University, evaluated what happened in a test restaurant where diners were offered a free glass of wine. In one instance, the wine came from a bottle with a fancy, graphically appealing label indicating it was from Sonoma, California. In another, the complimentary glass of wine came from a bottle with a very plain label, indicating it was bottled in Fargo, North Dakota. When the diners believed they were drinking the wine from California, they ranked the wine and the meal they ate with the wine better than diners who had the same meal with a bottle presumed to be from North Dakota. The wines were identical; both free glasses of wine were the same "Two Buck Chuck" purchased from Trader Joe's. The diners' reactions to the labels and where they believed the wine came from affected their satisfaction not only with the wine but also with the meal they ate with it. In short, even their taste buds were "hijacked" by the impression made by the wine label. Winemakers and restaurateurs from across the world sat up and took notice.

In another study, researchers at Stanford School of Business evaluated how consumers made choices or purchases that they believed would make them feel happy. This has become a critical strategy for businesses that try to develop brands that consumers believe will provide or enhance happiness. In fact, many examples of the advertising strategies are rather obvious:

> Nesquik claims, "You can't buy happiness, but you can drink it." Dunkin' Donuts promotes a breakfast sandwich as "The happiest sandwich on Earth." Nivea offers a body lotion, "Happy Sensation." Hugo Boss offers "Orange, the fragrance of happiness," and Clinique similarly offers a perfume named "Clinique Happy." Through interactive campaigns, marketers have also sought to cultivate happiness. Coca-Cola launched the "Open Happiness" campaign, which recognizes life's simple pleasures and encourages consumers to take a small break from the day to connect and share happiness with others. BMW developed a "Stories of Joy" global communication campaign that hosts consumer-created videos highlighting the joy of driving. Whiskas [cat food] encourages consumers to share their "Happiness with Whiskas" cat moments and become a member of the "Happy Together" online community for feline lovers.

One of the reasons advertisers and manufacturers spend so much effort promising to create a state of happiness for their consumers is that it has a profound effect on the purchases they make. To the extent that consumers create a reliable linkage between a product and feeling happiness, it forms the basis of "brand loyalty."

Marketing Happiness: Happiness as a Habit

If we don't know how to navigate the path to happiness, advertisers and marketers do. They have at their disposal the most remarkable tools for dissecting our appetites, weaknesses, and aspirations. We are naive to believe that we are not transparent to the sophisticated analyses being carried out by the likes of Apple, Google, and Yahoo. One hundred and twenty billion dollars is no joke, and that is how much is being spent on online advertising every year. The trick is that if someone, an agency, a corporation, has the ability to track every keystroke a person makes, every place their eyes dart, or where their fingertips hover, hesitate, and move on using cookies, ETags, and web beacons, *they have gained direct access to how that person thinks.* This means it is easy to figure out what worries him. What interests her. What tempts him. What does she dream about? Aspire to be?

The average American is spending approximately eleven and a half hours a day on a computer. Given that the average American also sleeps about seven and a half hours a day, that means that approximately 70 percent of each waking day is spent on a computer and every minute of that is available for analysis. For example, parents-to-be are every retailer's dream because they are consumers who must buy goods—lots of them—and goods they have never consumed before. So, what retailers see is a dramatic change in shopping routines and purchase history. It is a high priority for marketers to know—to deduce—by your keystrokes if you are pregnant. How could they tell?

There are a hundred clues. Checking out pregnancy tests. Looking up ob-gyn specialists or midwives. Buying maternity clothes. Blue paint. Pink paint. Strollers. Looking up names for boy or girl babies. Applications for maternity leave. It's a trail as wide as the Grand Canyon, and the marketers are standing ready to rush into that gap and sell merchandise to the expectant couple. That goes for the advertiser who gets there first with the best targeting for what you need, whether it be the Rolls-Royce of strollers—the Silver Cross Balmoral that goes for $4,000—or a used stroller on Craigslist that can go for less than $50. It also turns out that expectant mothers turn up their buying powers when they are in their second trimester, so advertisers target them with special buying accounts, newsletters ("the 12 must-have items for pregnancy"), and targeted advertising ("your trimester-by-trimester checklist").

Target, the American retail corporation, employs a staff of mathematicians who used artificial intelligence to develop an algorithm based on key product purchases that created a very accurate "pregnancy prediction score." So accurate, in fact, that one father went into his local Target store demanding to speak with the manager because his daughter had started receiving materials aimed at pregnant mothers. He accused the store of encouraging his teenage daughter to be sexually active and get pregnant. Later, the father called the manager and apologized: his daughter informed him that she was, in fact, pregnant and had been trying to hide the fact from her dad. Apparently, you can fool a parent but not the mathematicians at Target. As one Target marketer put it, "Just wait. We'll be sending you coupons for things you want before you even know you want them."

In the meantime, every time you, as a pregnant mother, turn on your computer and there is a spot on your screen to park an ad, you can bet it will be a crib, a breast pump, or a diaper bag. And you will never see a Dodge Ram truck there. But your husband will. And he'll also see ads for power tools since he has been thinking about building the addition for the nursery, but he will also need somewhere—a pickup truck, maybe?—to throw all those tools in the back. Learning your buyers' habits pays off big time because, according to one Duke University study, up to 45 percent of purchases are based on habit. *Analysis of our keystrokes and use of search engines allows marketing specialists to understand what motivates our personal purchases of retail goods.*

Looking for Happiness

In his book *Orbiting the Giant Hairball,* Gordon Mackenzie, a creativity guru and activist, points out how something has gone awry in our educational system. He can ask first graders, "How many artists are here?" and 70 percent of the students put their hands up. By the sixth grade, less than one in ten will identify themselves as artists. So, where did all the artists go? Our educational system killed off the creative confidence the first graders initially brought to school with them.

Suppose 50 percent of the happiness set point is genetically predetermined. In that case, we cannot afford to let our educational enterprise kill off the ingredients that make up the other 50 percent. As schools are increasingly driven by political agendas to demonstrate high scores in reading, writing, and mathematics, we miss the point that the children in our schools are not robots being programmed only for success in science, technology, engineering, and math. Our schools should be on a quest to find out what subject areas, experiences, and talents can be found that will help individual students feel happiest. It needs to be a long view of happiness as the cultivation of a personalized, lifetime pursuit. *Educational institutions often quash children's native curiosity and talents in the name of teaching them to pass standardized tests.*

9

On the Nature of Laughter

The Greatest Laugh Ever Heard

I can say categorically that the times in my life when I have been happiest have been moments of laughter. I can also identify the happiest laugh I have ever heard: my grandson's. He laughs when I try to get his shoes on. He yanks his foot back so both his sneaker and his sock come off all at once. I must have put that sock and sneaker back on a dozen times. Every time, he waits until I get his foot all the way into the sneaker, and then he yanks his whole leg back, leaving me with a feigned look of surprised and dismay that once again he had literally eluded my grasp. Gosh, it felt glorious. It is as if he has figured out how to inject my soul with heroin equivalents of happiness. I know. I just said scientists weren't absolutely sure we know what happiness really is. Okay, sue me. Call it joy then.

I remember thinking, I hope this is what heaven will sound like. And, while I confess that there may still be some minor scientific disagreement as to verifying that my grandson does, indeed, have the funniest laugh in history, all parents (and grandparents) reaffirm this tenet: there is no laugh that sounds as wholesomely delicious, carefree, and spiritually uplifting as that of a child.

A while back, I went with my whole family to take my youngest grandchild, Jake, to see the very first movie in his life in a movie theater: *Frozen II*. Every member in my immediate family—we were two children and eight adults—was on hand to witness the event. I kept asking myself, is it that critical that we all bear witness to him losing his Dolby-surround-sound-enhanced, cinematic virginity? But then I realized what our real motivation was: we all got to watch and hear him laugh at Olaf, the wonderfully funny, meltable, lovable, and singularly un-insightful snowman in the animated film series. We would hear Jake laugh out loud at Olaf, and then eight adults would laugh back in unison at his peals of laughter. Every one of us walked out of that movie theater with the biggest grins on our faces. Yes! Our (his) first cinematic outing had proven a great success because Jake had laughed so hard and so often! I kept asking myself, when did I lose the gift of laughing like that? And what a blessing a laugh bestows on everyone who can share it.

Laughter is telling us something very important about the nature of the brain. But exactly what has been difficult to discern scientifically; it turns out the science of laughter is not a simple matter. As ubiquitous as laughter may be, it is mysterious. Laughter has been broken down into several categories: social laughter, fake laughter, genuine laughter, induced laughter, and pathological laughter. The writer E. B. White wrote, "Analyzing humor is like dissecting a frog. Few people are interested, and the frog dies of it." William James, the American father of modern psychology, put his finger on the enigma that is laughter more than a century ago. Early in his career, he had carried out some experiments with inhaling nitrous oxide, also known as laughing gas. It led him to what I consider one of the most profound and elementary insights about the nature of laughter. James wrote, "We don't laugh because we are happy. We are happy because we laugh."

Only Primates Can Laugh

The current explanations we have for laughter are neurologically scant and scientifically inadequate. Still, laughter is such an important ingredient in our lives that I want to do my best to tease out as much as science will currently let us. Laughter is a rare gift, given to few other species in the animal kingdom. It appears to have evolved over the last ten to sixteen million years of evolution and exists only among primates (see figure 9). Tickle-induced laughter is observed only among orangutans, gorillas, chimpanzees, bonobos, and humans. *Only five species of primates share this rarefied trait of tickle-induced vocalizations.* Chimps, for example, often use tickling during grooming as a mechanism to enhance bonding within the troop. The vocalizations, along with an expansive, relaxed smile, reassure companions this close form of socialization is safe. Laughter serves much the same role in humans, enhancing cohesion in the way singing, dancing, and playing music together do.

Of the five primate species, however, only *Homo sapiens* evolved a unique form of *social* laughter. We all use it frequently—as many as twenty times a day on average. Women tend to laugh less as they get older. Men don't. Why? No idea. Maybe men just aren't mature enough to realize that there's less to laugh about as they age. Human beings tend to laugh more toward the end of the day—a good reason to avoid any sunrise comedy acts. Whenever I wade into the subject of humor and laughter, I am aware it can quickly carry me into the realms of philosophy and spirituality. It's a mercurial subject. Aristotle asserted that laughter was unique to human beings. Wrong. He also believed that babies had no soul until they started to laugh and that it always happened on the fortieth day after birth. Wrong again. Nietzsche maintained laughter was the result of existential loneliness. Naturally. And Freud maintained that laughter primarily occurs as a mechanism for relief of anxiety. Of course he did.

Don't Laugh at Powerful People

Neither Aristotle nor Plato trusted laughter because they asserted it undermined authority. And they were right: nothing undermines someone's authority faster than becoming the subject of derision and ridicule. But that is really just one useful side effect of laughter, not its point. The truth is it is hard to be fearful or in awe of someone in power when you can laugh at him. I suspect it is this abiding truth that kept President Trump from attending the annual White House Correspondents' Association (WHCA) dinner throughout the four years of his administration. The WHCA has held an annual dinner for the president since 1921. It is a venerated mainstay of Washington, D.C., societal and political tradition.

President Trump, however, never attended a single WHCA dinner—something no other full-term president has ever done. He issued an order that no cabinet officers or representatives of the government attend the dinner either—again, something no other president has done. At one point, President Trump hinted that he might be willing to attend a WHCA annual banquet on one condition: that a comedian not be the featured speaker. Aristotle and Plato were right to be concerned. Trump understood that humor was his kryptonite.

Charlie Chaplin savagely parodied Hitler on the silver screen in his movie *The Great Dictator.* Hitler was something of a movie buff, and it was rumored that he had a print of the movie smuggled into Germany. He was reportedly so enraged when he saw it that he said he wanted to see Chaplin dead, but no official orders to kill the actor were ever issued. Hitler did, however, formally ban the movie from ever being shown anywhere in the Third Reich. Oh, how the mighty fear the power of laughter.

As a vocalization, human laughter has been studied extensively. It is typically a series of short notes (seventy-five to a hundred milliseconds long) that occur in rapid succession with about two hundred milliseconds between notes that all take place during an exhalation. This is stereotypical and acts as a social emollient, as it were. I have often gone visiting to countries as a lecturer where I did not speak the local tongue and where I found myself stuck with someone who did not speak a word of English. We often spent most of our time smiling at each other and creating a context where we could use social laughter to defuse the predicament. Laughter has a temporary ability to supersede whatever emotions are ruling the day, be it shyness, embarrassment, or discomfort. So, a laugh is a good way to break the ice.

I get social laughter and how we use it—being the sophisticated chimps we are. What I don't believe science has figured out is the nature of *real laughter.* The involuntary laugh. We all know a fake laugh, like the one we give when someone makes a lame joke and we are laughing just to be polite. It is forced and mediated by the frontal lobe. Scientists call it a "cognitive laugh." Anyone who hears it knows right away it is a fake. But the other laugh, that uncontrolled, ready-to-fall-to-the-floor, practically peeing-in-your-pants kind of laugh, that is mediated through the cerebellum and is completely reflexive in nature. It arises from the ancient, reptilian part of our brain and is beyond our conscious control. Its authenticity is unquestionable. These cerebellar laughs still have interesting social cues embedded in them.

For example, we are far less likely to have a cerebellar laugh when we are alone. We like to laugh when we are with others. Cerebellar laughter is also socially contagious. When someone (or a few individuals) within a larger group starts to chuckle, you can practically see it starting to hop from one person to the next. Many is the meeting where I have had to secretly bite the inside of my cheek to try to keep a straight face, hoping the pain would keep me from chuckling aloud.

The Theories About Laughter

Some have theorized that mirror neurons may underlie contagious laughter. In 1962, there was an outbreak of contagious laughter of epidemic proportions. It became known as the Lake Victoria or the Tanganyika laughter epidemic. It started among a handful of girls at a boarding school, but it then spread inexplicably from one group of schoolchildren to the next and kept reinforcing itself. It spread across the country and spilled over into neighboring Uganda. Teachers and administrators could get nothing accomplished with the students laughing hysterically. Eventually, dozens of schools were shut down for several weeks, and the students were sent back to their villages. Once the children were housed in their separate villages and dwellings, the outbreak finally subsided over a period of months.

Few things can make us feel as happy as quickly as laughter. Whenever it appears, it takes over. As Stephen King wrote, "You can't deny laughter when it comes, it plops down . . . and stays as long as it wants." It seems to monopolize everyone and commands everyone's attention. It is such a tonic for whatever ails you. Laughter feels so good because it stimulates the release of endogenous opioids. That laughter-induced secretion of endorphins, in turn, can entrain more laughter by contagion, which, in turn, engenders more endorphins. Thus, it can give us an upward spiral of laughter much like what we saw with oxytocin in romance.

There's also the International Society for Humor Studies that helps host international conferences with presentations titled "Did Hitler Have a Sense of Humor?" and "Humor and Scatology in Contemporary Zulu Ceremonial Songs." And, of course, "Sex and Women in Biblical Narrative Humor." Plato and Aristotle believed that humor was supported by a notion of feeling superior and arose from looking at the misfortunes of others lesser than oneself. Freud, as I mentioned earlier, argued for a relief theory—namely, that humor served as a psychological mechanism to overcome social anxieties and inhibitions. But most comedic theorists believe that humor arises from inconsistency—what is called incongruity theory. Humor arises when there is some alteration between the course of events that people expect to see and then what actually ends up happening. Certainly, that is the case with most jokes that have a punch line. The punch line takes us somewhere we did not expect to be carried by the setup of the joke. And that may explain punch lines, but it does not tell us why joke itself is funny.

The latest theory, which is offered by Peter McGraw, a researcher at the Humor Research Lab (abbreviated as HuRL) at the University of Colorado, is what he calls the "benign violation theory." Henri Bergson noted that in order to laugh at, say, Harold Lloyd swinging precariously in the air from a clock and struggling in panic as the clock begins to come apart under his weight, one needs to feel indifference to his plight. The same goes for watching someone slip on a banana peel. You laugh first and then stop yourself because you should not be laughing, because they might have been seriously injured in the process.

Bergson wrote,

> Indifference is . . . [comedy's] natural environment, for laughter has no greater foe than emotion. I do not mean that we could not laugh at a person who inspires us with pity, for instance, or even with affection, but in such a case we must, for the moment, put our affection out of court and impose silence upon our pity.

So, something is humorous when something happens that shouldn't—and that could be dangerous (or even lethal)—but eventually ends up being harmless? I'm beginning to see what White meant about dissecting a frog. However, brain imaging studies looked at what made professional comedians different from amateur comedians or a control group (presumably, made up of *un*funny people), and the answer appears to be activity in the temporal lobe. It would appear that comedians rely heavily on their own experiences and memory in the hippocampus to draw parallels between, for example, what happened and what would have

Harold Lloyd hangs
precariously in the silent
film *Safety Last!* (1923).

been expected to happen under normal circumstances.

Ori Amir, an assistant professor of psychological science at Pomona College (and a stand-up comedian in his own right), evaluated comedic humor across stand-up acts, written jokes, cartoons, and cartoon captions and noticed that good humor usually has a highly creative quality to it. It presents an original perspective or an unexpected outcome. The *Far Side* cartoon anthology by Gary Larson is a perfect example of how a uniquely personal, quirky creative take on monsters, aliens, and the impact of science produces laugh-out-loud funny cartoons, often barbed with a sharp point to make.

One of my favorite cartoons was one produced by the cartoonist John Callahan. It depicts a typical sheriff's posse on horseback tracking someone down and coming upon an abandoned, overturned wheelchair in the desert. One of the men in the posse says, "Don't worry, he won't get far on foot." Of course, the joke is in terrible taste. But its humor arises from the fact the posse is in earnest tracking a suspect who is paralyzed, confined to a wheelchair (which he has abandoned), so he *really* isn't going to get far, because he must drag himself bodily along the ground to escape.

Callahan himself had been rendered paralyzed in a car accident and spent the last three decades of his life in a wheelchair, specializing in tasteless jokes about people with disabilities. So, why is the cartoon funny? Because we feel uncomfortable about disabled individuals being confined to a wheelchair. It troubles us. The cartoon makes the situation even worse: a man in a wheelchair is being hunted down by a posse on horseback! It underscores the impact of the disability. We expect some measure of pity for the man being pursued. Not elation that he is about to be apprehended by the posse.

Likewise, Callahan had also made a cartoon of a cliff with a sign that read, "Suicide Leap." Next to the sign there was a ramp for wheelchair access and a "Handicapped" sign. Again, the cartoon tries to resolve the incongruity of ensuring disability access so one can hurl oneself off a cliff. In these cartoons, Callahan harnessed his unique insight into the anxiety that disabilities can create in a world of able-bodied individuals. As Scott Weems wrote in his book *Ha! The Science of When We Laugh and Why*, "We feel sorry for people with handicaps, but we also want to empower them and treat them as they should be treated—like everyone else."

What we now know from looking at brain scans of people as they look at cartoons and laugh is that the humorous reactions are accompanied by activity in the amygdala followed by hyperactivity in the ventral tegmental area (VTA). So, the brain has literally changed the outcome from a situation where we initially experienced anxiety and stress—as demonstrated by activation of the amygdala—and resolved it into a pleasurable, satisfying response with an outpouring of dopamine from the VTA. So, neuroimaging would suggest that resolution plays a big role in humor.

Laughing at Funerals

That brings me to an important topic I need to touch on: funerals. Death—the business of funerals—is obviously not a laughing matter. Or is it? I would like to think that my experience with people passing and the funerals that follow is about average. But I must share with you that there are good funerals and bad ones. Granted, people do not usually go around penning a daily column with their reviews and analyses of funerals, as they do of plays opening on Broadway. But if we—okay, I—if I were to be a funeral reviewer, say, for *The New York Times,* I would advocate for funny funerals. They would receive the coveted five-star review because the best eulogies are the ones that make people laugh.

I am sure laughter is so cherished at eulogies precisely because it helps relieve the almost unbearable anguish we feel about losing someone we love. A while back, my family lost a beloved matriarch named Rita. At her funeral, one of her son's (with a wonderful sense of humor) stood up at the lectern and began his eulogy, "For those of you who did not know Rita, you're too late," and gestured toward the coffin. Clive James stated, "Common sense and a sense of humor are the same thing, moving at different speeds. A sense of humor is just common sense, dancing." I like to think we were symbolically dancing at Rita's funeral by sharing all the laughs she had shared with us.

Laughter Is the Best Medicine: Norman Cousins

Norman Cousins (1915–1990) is one of the heroes who lies in my personal medical pantheon. He was a celebrated writer and legendary editor at the *Saturday Review* magazine. He had several health issues to contend with, including significant atherosclerotic heart disease and crippling ankylosing spondylitis (AS). AS is believed to be a disease that is triggered by dysfunction in the immune system and leads to a systemic and painful inflammatory reaction that causes the vertebral bodies, the bony building blocks of the spine, to fuse together, gradually restricting movement of the spine and producing painful compression of the neural elements and the spinal cord itself. The disease can become so severe that the ribs of the thoracic cage can fuse together and gradually restrict breathing. There is no known cure for AS. Mr. Cousins was confined to bed and went on leave from his editorial position at the *Saturday Review.*[*] His doctor had informed him that he had a less than one in five hundred chance of improving and that he was likely to die within six months.

Mr. Cousins decided not to take the news lying down, literally. He discharged himself from the hospital. He felt the environment was toxic and, instead, checked into a hotel. He put himself on massive doses of vitamin

---

[*] I reserve the right to call people by whatever title best fits my comfort. I had the opportunity to meet Mr. Cousins and could not imagine calling him anything but Mr. Cousins.

C and a rigorous organic diet. And he purchased a movie projector. He then launched himself on a do-it-yourself behavioral modification project. He obtained movies of Buster Keaton, Laurel and Hardy, Charlie Chaplin, the Marx Brothers, episodes of *Candid Camera*. In short, he started viewing anything and everything that he was sure would make him laugh—belly laugh—hard enough that it hurt. He laughed through the pain of his AS. Mr. Cousins became convinced that he was slowly getting better. He reported, "I made the joyous discovery that ten minutes of genuine belly laughter had an anesthetic effect and would give me at least two hours of pain-free sleep. When the pain-killing effect of the laughter wore off, [I] would switch on the motion-picture projector again, and, not infrequently, it would lead to another pain-free sleep interval." In six months, he was able to ambulate and in two years returned to his full editorial duties at the *Saturday Review*.

Mr. Cousins lectured widely and wrote a best-selling book, *Anatomy of an Illness as Perceived by the Patient*. Mr. Cousins's remission baffled medical experts. There was no question that he was markedly improved. He used his position to promulgate for a new style of integrative medicine that included a more positive attitude toward the patient's clinical improvement. He helped to launch research efforts to better understand how emotions were linked to biochemical changes. He enunciated the vital role that physicians should play in harnessing a patient's psycho-neuroimmunological responses and described the best doctors as being

> not just superb diagnosticians but men who understand the phenomenal energy (and therefore curative propensity) that flow out of an individual's capacity to retain an optimistic belief and attitude toward problems and human affairs in general. It is a perversion of rationalism to argue that words like "hope," or "faith," or "love," or "grace," and "laughter" are without physiological significance. . . . [B]enevolent emotions are necessary not just because they are pleasant but because they are regenerative.

Mr. Cousins proved his doctors wrong and proved his point: he would eventually die from heart disease sixteen years after his AS was diagnosed.

Laughter produces a dramatic, virtually instantaneous change of mood in people. Gelotology (yes, there is a technical name for laughter research) has shown that laughter is a universal language. It is seen across cultures and ethnic groups. It is universally understood and appreciated. Babies can laugh long before they learn to talk. The healthful side effects of laughter have begun to be elucidated. Lee Berk, a psychoneuroimmunologist at Loma Linda University, evaluated subjects who were shown an hour-long humorous video. Blood samples were taken for analysis at ten minutes before, thirty minutes into the viewing of the tape, and thirty minutes and then twelve hours after viewing the video. The blood samples from the laughing test subjects demonstrated significant increases in the activity levels of more than six different subtypes of lymphocytes, cells responsible for mediating our immune response. They included active cytotoxic T cells, natural killer T cells, and helper T cells, which are responsible for enhancing and modulating our immune responses. There were also statistically significant increases in circulating IgG and IgM (antibodies) and cytokine interferon γ, all of which activate and then direct white cells where they are needed to fight off invading microorganisms. In other words, one short bout of watching a humorous movie and laughing primed the entire immune system for a sustained period of more than twelve hours! Psychoneuroimmunological studies reveal that laughter is associated with elevated levels of endorphins and enhanced immune function.

Laughter has been shown to be a useful adjunct in cardiac procedures and rehabilitation. Post-myocardial-infarction patients were provided with identical cardiac rehabilitation programs, but one group was allowed to view self-selected humor for thirty minutes a day. During the subsequent year after suffering a heart attack, the humor-self-treatment group suffered fewer episodes of arrhythmias and had lower blood pressure, lower circulating catecholamines (which can cause cardiac stress), lower requirements for adjunct cardiac medications, and fewer subsequent heart attacks. Mirthful laughter's ability to diminish stress levels appears to work through three different mechanisms. One is a direct effect of reducing the levels of circulating catecholamines, and hence it lowers sympathetic activation. The second mechanism appears to be from activation of the hypothalamic-pituitary-adrenal axis, by promoting the release of growth hormone, prolactin, and opiopeptides. The third appears to be the ability of circulating neurochemicals to induce an enhanced defensive immune response.

Experiments have also confirmed that laughter diminishes the perception of pain, presumably from the release of enkephalins and endorphins. Recent experiments in Scotland on laughter's effect on pain perception used clips from *South Park, The Simpsons,* and *Friends* and clips from a number of stand-up comedians. In these same experiments, investigators evaluated whether so-called feel-good videos like *Planet Earth* were comparable in their effect to humorous videos. Only the funny videos were able to induce analgesia. So, next time you are arguing over the television remote, remember comedy rules.

The striking thing about Mr. Cousins was how his intuition about his own medical predicament led him to make some remarkable deductions about the nature of laughter. Its ability to relieve pain. Its ability to heal. To restore health and purpose. Mr. Cousins understood the magic in laughter long before we understood the science behind it. Suppposedly, the late, great science writer Isaac Asimov said: "The most exciting phrase to hear in science, the one that heralds new discoveries, is not 'Eureka!' (I found it!) but 'That's funny . . .'" The funny thing is that it is absolutely true.

10

On the Nature of Stress, Fear, and PTSD

Monkeys? You think a monkey knows he's sittin' on top of a rocket
that might explode. These astronaut boys know that, see? Well.
I'll tell you something, it takes a special kind of man to volunteer
for a suicide mission, especially one that's on TV.
—Chuck Yeager's character in the film *The Right Stuff*

At one time or another, stress will get the better of us. Most of us
can recall a school exam where we knew the material cold, but
under the pressure of taking the test, we simply couldn't recall the
answers. Anyone can belt out a tune in the shower, but try singing the
same song in front of ten thousand people. Stress can completely unravel
our ability to perform.

Stress: A Little Goes a Long Way

The Yerkes-Dodson Curve

In 1908, two psychologists, Robert Yerkes and John Dodson, investigated
how stress affected a rat's ability to negotiate a maze. They found that the
neuroendocrine system of each rat had a unique set point, where the rat
was able to perform simple and complex visual tasks in puzzling out how
to get through the maze to find the reward of a food pellet at the end.
The researchers began administering a series of shocks that increased
incrementally in strength while the rats negotiated the maze using the
visual cues. At first, small shocks motivated the rats to run the maze more

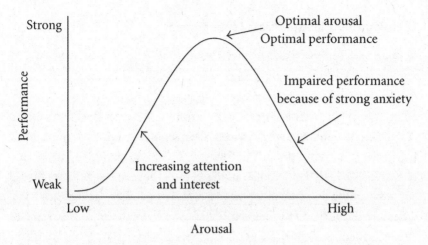

The Yerkes-Dodson curve is often called the bathtub curve because of its inverted U shape. It demonstrates the principle that a low level of stress enhances our concentration and can help improve our performance, but it will reach a "sweet spot" where our performance has been maximized. However, if stress is increased beyond this tipping point, then it becomes deleterious and will degrade performance. Each of us has our own unique "sweet spots" in our makeup where a little stress helps improve outcomes and beyond which stress becomes destructive.

quickly. So, low-level stress improved the rat's learning ability.

However, as the shocks became increasingly painful and stressful, the performance began to deteriorate. So, the data yielded the so-called Yerkes-Dodson law, often referred to as the bathtub curve because it resembles an inverted tub. It shows that small amounts of arousal will maximize performance; however, if the arousal is increased to the point it triggers alarm and anxiety, learning success decays quickly and dramatically. Small increments of stress can help improve our performance, but large amounts of stress will cause us to fail.

The Yerkes-Dodson curve also predicts individuals will have a set point where attendant stress will become too high and their performance will break down. But that also begs the question: What happens if a person's performance cannot be permitted to fail? For example, is a pilot allowed to break down under stress and fail to perform in the face of a possible impending crash? Or is a Secret Service agent allowed to hesitate under stress and fail to shield the president?

Learning the Ultimate Stress Management

Sullenberger and the "Miracle on the Hudson"

On January 15, 2009, US Airways Flight 1549 took off from LaGuardia Airport en route to Charlotte, North Carolina. The aircraft was an Airbus A320-214, powered by two massive GE/Snecma engines, each producing thirty-four thousand pounds of thrust. During the plane's initial ascent, it passed through a flock of Canada geese. These are very large birds, some weighing up to twenty pounds, and several were sucked into both jet engines. The plane suddenly lost all power before even reaching three thousand feet of altitude. The pilot, Captain Chesley B. "Sully" Sullenberger, was fifty-seven years old. He was a former fighter jock and had logged more than twenty thousand hours as a pilot. Captain Sullenberger immediately notified air traffic control that they had an emergency with loss of power in both engines. Less than six minutes after he took off, Captain Sullenberger had to glide the jet without power to a water landing on the Hudson River. He purposefully picked a site downriver, adjacent to the Intrepid Museum and the Circle Line cruise terminals, to maximize the number of watercraft that could help rescue any survivors. However, not a single life was lost. Later, a National Transportation Safety Board pronounced it "the most successful ditch in aviation history."

Secret Service

Would a Secret Service agent be forgiven for being so overwhelmed by the stress of encountering an unexpected assailant that he failed to shield the president? How do individuals prepare for such mission-critical jobs? For agents in the Secret Service, there are grueling written evaluations, extensive background investigations, psychological testing, and physical fitness qualifications. Each agent must pass all on the first try. There are no redos.

In addition to the usual training regimen, all Secret Service agents must spend a few years in the agency's investigative arm before being considered for additional training required for assignment to a protection detail. The protection details have their own grueling standards. One has only to be reminded of the behavior of the Secret Service agent Clint

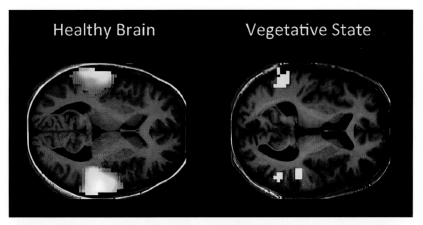

Figure 1. Functional MRI showing significant difference between a healthy brain (A) responding to an auditory stimulus versus a patient in a vegetative state (B); today, it is called unresponsive wakefulness syndrome. There is much more activity in the auditory area in the healthy brain and virtually no response in the individual labeled as being in the vegetative state. *Photo is courtesy of Dr. Adrian Owen, OBE, and is reproduced with the author's permission.*

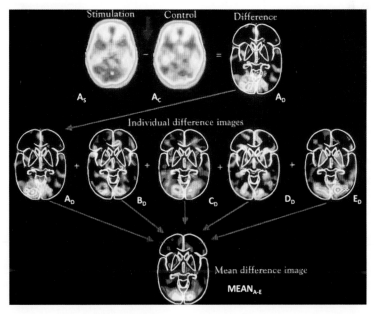

Figure 2. Functional MRI explained: Images of individual A are obtained at rest and serve as baseline control resting state (AC). Then we obtain images in the scanner (*large red arrow*) while the brain is being stimulated (AS). Taking the activity in AS minus that seen in AC gives us the difference between the two images, labeled AD. We could also average the differences seen on five subjects (A–E, for example); then we obtain the average (or mean) difference of the group's responses (MEANA–E). This allows us to compare the average brain activity between different groups of individuals. *Permission to use photo courtesy of* Science.

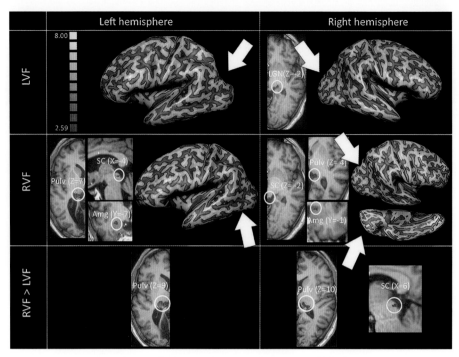

Figure 3. Functional MRI of subjects who suffer from cortical blindness who are being shown a photograph of someone with an angry expression. Even though the subject is blind, we can see his or her brain has activated a purely *subconscious* emotional response to the facial expression captured in the photo. Areas in orange show increased activity in regions of the brain that interpret emotion and regions in the midbrain that subconsciously intercept visual signals. These images show us that our brains can subconsciously "see" even when we are consciously blind. The color bar on the top left reflects values of the respective brain activation contrasts. *Permission to use figure courtesy of the* Proceedings of the National Academy of Sciences.

Figure 4. *Panel above:* A view of our planet Earth (bracketed between the two white lines) taken by *Voyager 1* from a distance of more than 3.5 billion miles away. At this distance, even in *Voyager 1*'s narrow-angle camera, Earth is only 0.12 pixels in size. *Panel below:* An enlargement of the "pale blue dot" (bright spot at the center of photograph) that is our home planet Earth. *Photograph courtesy NASA/JPL and the* Voyager 1 *spacecraft.*

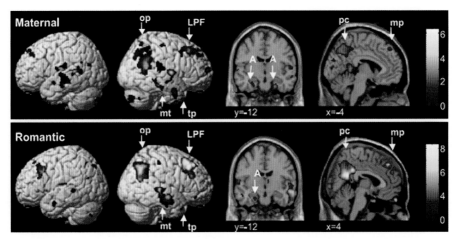

Figure 5. A series of fMRI scans reviewing activity in a mother's brain as she looks at pictures of her child (*upper row*) versus when she looks at pictures of her partner (*lower row*). There is a striking concordance between parental and romantic love. In other words, becoming a parent is akin to falling in love. *Image is courtesy of* NeuroImage *and the authors Bartels and Zeki,* NeuroImage *21, no. 3 (March 2004): 1155–66.*

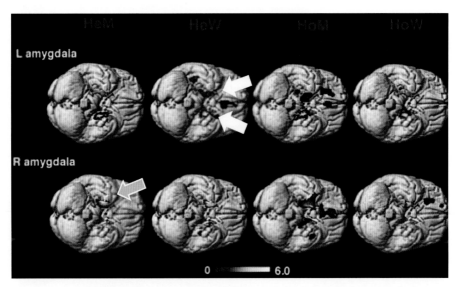

Figure 6. An fMRI study demonstrating differences in brain activity (orange color) are linked to sexual preference. The right amygdala (*gray arrow*) in heterosexual men (HeM) and homosexual women (HoW) shows similar activity. This is in contrast to bilateral amygdala activation (*white arrows*) in homosexual men (HoM) and heterosexual women (HeW). Sexual preference for males produces similar brain activity in heterosexual women and homosexual men, while attraction to women produces similar patterns of brain activity In homosexual women and heterosexual men. *Permission to use the image courtesy of the* Proceedings of the National Academy of Sciences. *© 2008 National Academy of Sciences, U.S.A.*

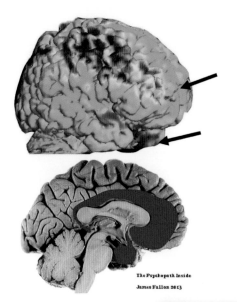

The Psychopath Inside
James Fallon 2013

Figure 7. *Left:* Brain PET scans from a group of psychopathic subjects. Left-hand panel shows diminished activity in the frontal lobe in green (*upper black arrow*) and the temporal lobe in blue (*lower black arrow*). *Right-hand panel:* Brain PET scan of a normal individual (upper images) and the brain of a psychopath (lower images). Note the much darker areas of the psychopathic scan (*white arrows*), showing diminished activity in the prefrontal and temporal lobes compared with control. *Permission to use images courtesy of James Fallon, PhD.*

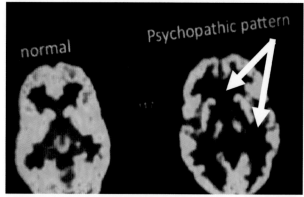

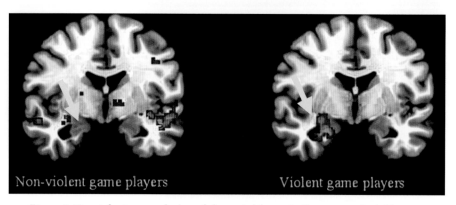

Figure 8. Nonviolent game-playing adolescents (shown in the scan on the left) showed much lower levels of amygdala activity (shown with yellow arrows) than violent game-playing adolescents (shown on the right-hand side). Permission to use figure from Radiological Society of North America.

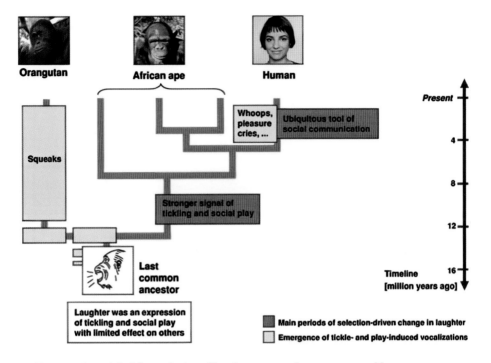

Figure 9. A model of the evolution of laughter among the great apes and humans.
Permission to use figure courtesy of Dr. Marina Davila-Ross.

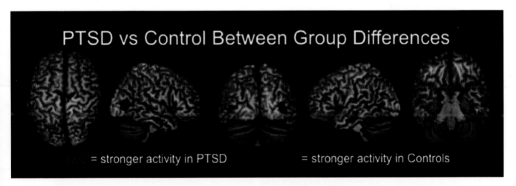

Figure 10. A 3-D rendition showing brain regions with significant differences in
the resting state activity in combat veterans with PTSD compared with a control
group of combat veterans without PTSD. Veterans with PTSD show much greater
activity (shown in red on the two images on the left-hand side) in the temporal lobes
(amygdalae, parahippocampus, and so on) that govern arousal and aggression.
In short, their brains are "amped" up with a sense of threat. *Copyright © 2017 Badura-
Brack, Heinrichs-Graham, McDermott, Becker, Ryan, Khanna, and Wilson.*

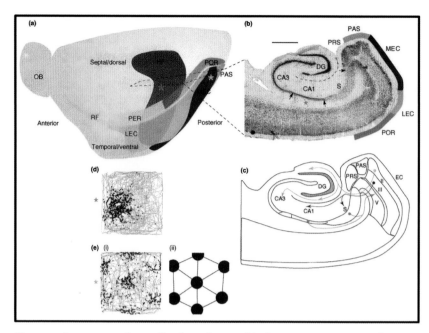

Figure 11. A summary of recording from the entorhinal cortex as a mouse wanders through a maze. Frame labeled *d* shows position cells that will fire when the mouse finds itself in one unique location (for example, "this is *the entrance* to *this* maze"). It will fire only when the mouse is standing in that single location. Frame *e* shows grid cells in their typical hexagonal arrays. Grid cells can fire many times in different locations and appear to remind the mouse that one location may closely resemble another (for example, "*this entrance* looks just like the *same kind of entrance we have seen in other mazes*").
Permission to use figure courtesy of Menno P. Witter, Edvard Moser, and Trends in Neurosciences.

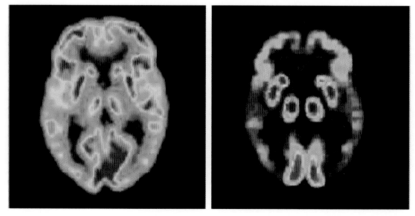

Figure 12. A PET scan of a normal volunteer on the left compared with an individual suffering from advanced Alzheimer's disease on the right. The brighter yellows and oranges indicate high activity (*red arrow*), while the blues and purples indicate much lower activity (*blue arrow*). The image of Alzheimer's on the right shows a dramatic global reduction of brain activity. These scans are courtesy of the National Institutes of Health and are in the public domain.

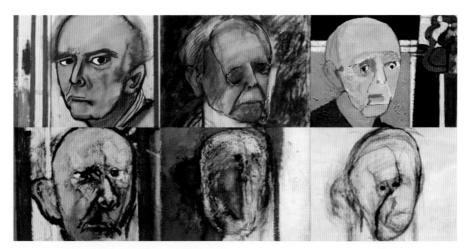

Figure 13. A series of self-portraits created by the painter William Utermohlen between 1996 and 2000. Utermohlen was diagnosed with Alzheimer's disease in 1995 and could not paint after 2000 (he died in 2007). This dramatic series of paintings captures the artist's own perception of losing his identity and the very essence of his humanity over the course of the illness. It also speaks powerfully to the devastating effect of Alzheimer's disease as it gradually strips away memory. *Permission to use figure from Jennifer Norback Fine Art Projects.*

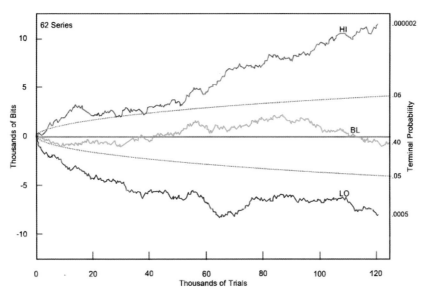

Figure 14. An example of the effects exerted by an individual over the course of nearly 400,000 trials on an RNG. A truly random effect of the RNG should correspond to the middle line (labeled *BL* for baseline). However, what was seen was that some individuals could make the RNG computer produce significant deviations away from the expected random output; that is, their minds altered the outcomes produced by computer. The upper line (labeled *HI*) and the lower line (labeled *LO*) represent just how dramatically the outcomes were made to veer away from random chance (*BL*). *The figure is courtesy of Roger Nelson, R. D and the PEAR laboratory.*

Figure 15. A photograph of Earth, known as *Earthrise,* taken by the astronaut William Anders on December 24, 1968, through the window of the Apollo 8 spacecraft. It has been called "one of the most influential images of all time." *This is made available by NASA and is in the public domain.*

Hill, who on November 22, 1963, jumped from one moving vehicle to another in Dallas so he could climb onto the back of President Kennedy's limousine and shield the fallen president and the First Lady with his body. Or Agent Tim McCarthy, who was struck by a bullet while using his body to shield President Reagan during John Hinckley Jr.'s attempted assassination in 1981.

For pilots, the key to staying focused during emergencies is to rely on their elaborate training regimens on sophisticated flight simulators. These are multimillion-dollar computerized machines that create highly accurate visual and tactile (or haptic) feedback that allow pilots to train extensively for all potential emergencies. Nowadays, all pilots and cockpit crew become certified on these high-fidelity flight simulators rather than on the actual aircraft itself because the simulators offer a much more exhaustive assessment. Malfunctions of every kind can be accurately simulated far better than they can on an actual aircraft. Sullenberger's hobby also happened to be flying gliders, so he routinely practiced how to handle an aircraft without power—years before he would need the skill to land a commercial jet. As one safety reviewer said, "Everything [Sully] had done over his professional piloting career contributed to his success here."

Likewise, in the James J. Rowley Secret Service training facility out-side Washington, D.C., agents train in full-sized mock-ups of the White House, Air Force One, Marine One, and the presidential motorcade. They use full-scale street scenarios to simulate potential attacks on the president. The kind of high-fidelity simulation training that airline pilots and Secret Service agents undergo produces what we refer to as a right-ward shift of the Yerkes-Dodson curve. Intensive, high-fidelity train-ing can helps us perform at high levels of success despite substantive increases in stress.

If we want to train someone's brain to remain in control, even in the face of stressful potential life-or-death scenarios, we need to first provide practice on a whole different level from what usually passes for training. The training must be followed up by sustained, repetitive, and frequent extreme practice exercises. This means there is no "now you've learned it and you're done" point. You're never done. And you're always trying to prove that you are able to maintain training at peak performance levels.

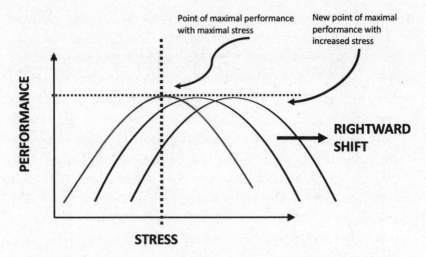

Repetitive training under stress can produce a rightward shift of the Yerkes-Dodson curve, allowing an individual to improve maximal performance under increasing levels of stress.

Second, as that practice becomes ingrained, the training environment must become increasingly high-fidelity and immersive. This means that with successive training, the practice environment will gradually become increasingly realistic until it is virtually identical to the real thing. Usually, this also means that the training environment becomes extremely expensive because we must create a training setup that is so realistic it employs every sensory signal—sight, sound, smell, the works—to try to "fool" the trainee's brain into reacting to everything in it as if it were real. This allows one to constantly "test" the ability of the frontal lobes to stay focused even in such a realistic format that it could pass for the real event. This means that if we want our Secret Service agents to keep their wits about them if a madman were to attack the president in the White House, then we have to practice in the White House—or a replica of it. And we need someone to realistically simulate how a madman might attempt to go after a simulated president with real drive and purpose. We also need to consider how such an attack might occur from every possible angle of approach. At every turn, we must look for how we can turn up the realism even higher.

This kind of sustained, high-fidelity practice allows the brain to solidify the memory of the requisite behaviors so they become second nature.

Our frontal lobes cannot be burdened with trying to puzzle out what is the right thing to do under stress. As I train my neurosurgery residents to handle life-threatening brain injuries, my adage is simple: "If you have to *think* about what's the right thing to do, then you don't *know* what's the right thing to do."

Third, high-level realism in simulation training creates all the sensory signals that would transmit stress to the individual. It "hardens" the frontal lobe, as it were, so that it recognizes all the cues in the life-threatening scenario, and since it has been able to handle them in the past (albeit simulated past), it will be able to handle them now. This produces the rightward shift of the Yerkes-Dodson curve.

We started by asking what we do with individuals in jobs or positions where they cannot be permitted to fail. The answer is train them so intensely and immersively that the rightward shift of the Yerkes-Dodson curve guarantees they won't fail under stress. These are the same three principles we used when I was in the military and trained special ops troops. These were rapid deployment units that would find themselves training for critical missions where they simply could not fail. As one of my sergeants summed up the training mission, "We train 'em hard, we train 'em long, and we train 'em deep, with every ounce of realism, because we cannot simulate the fog of war, so we have to make sure we have trained them well for everything else." That about sums it: *when training to encounter the unknown and the unexpected, train for everything else.*

Again, neuroimaging may begin to give us more tools in this regard. Recent research tells us how we can assay individuals under stress and identify those with the lowest levels of stress markers. These are the people who will be the least likely to be overwhelmed. Recent brain imaging studies in mice show that stressful situations caused them to release a neuropeptide called brain-derived neurotrophic factor (BDNF) in the nucleus accumbens. Mice that remain stressed will continue to show high levels of BDNF output, while rats that have adjusted well to the stress will have brought their BDNF back to baseline levels. In the future, we may be able to assess an individual's ability to stay coolheaded in stressful situations by assessing his or her BDNF response curves under increasing stress. Save the realistic, immersive training for them.

Perceived Versus Actual Threats

While we undoubtedly live in a much safer world than our ancestors did a hundred thousand years ago, several issues might *appear* to make our world more dangerous and create a greater sense of generalized anxiety. For example, I grew up in a low-to-middle-income working-class neighborhood. We walked about a mile to our elementary school, by ourselves. My mom only walked me there on the first day of school and on the last day when I graduated to go on to junior high. In the afternoons and on weekends, we played endlessly in a large open communal backyard. Baseball. Football. Hide-and-seek. Cowboys and Indians. We built forts. Ran bike races. Set up obstacle courses. We even ran soapbox derby races. On Halloween, all the kids roamed the neighborhood in our costumes in the company of either friends and peers or older siblings. Not once in all my childhood did I ever hear an adult caution us about the danger of strangers, the possibility of abduction, or molestation. It just was not on our parents' minds and, therefore, it was not on ours.

Five decades later, in that same neighborhood, kids travel that same distance to school but in a bus and under the watchful eye of a volunteer parent who accompanies the bus to the drop-off point on school grounds. The area where I and my friends roamed freely is now enclosed by a high chain-link fence encircling the playground. We made up all the games we played free from interference by adults. Now the only games being played there are organized Little League baseball games or soccer matches where everyone is in a team uniform and dads run up and down the sidelines with clipboards and whistles under the bright lights. Halloween is now strictly supervised, with assigned teams of parents as escorts, and the kids go to homes where the parents are well known to each other.

What happened? Did the world get suddenly more dangerous in four decades? No, quite the contrary. Childhood abductions and killings are almost half of what they were when I was growing up. But the media's ability to broadcast news stories about childhood homicides and kidnappings and to sensationalize them gives us the impression of growing, looming danger.

Bruce Schneier, who was a chief technology officer for IBM and blogs regularly on security, writes in his book *Beyond Fear,*

People exaggerate spectacular but rare risks and downplay common risks. They worry more about earthquakes than they do about slipping on the bathroom floor, even though the latter kills far more people than the former. Similarly, terrorism causes far more anxiety than common street crime, even though the latter claims many more lives. Many people believe that their children are at risk of being given poisoned candy by strangers at Halloween, even though there has been no documented case of this ever happening.

Daniel Gilbert, the author of *Stumbling on Happiness,* is a professor of psychology at what must be one of the happiest research quarters at Harvard University, the Hedonic Psychology Laboratory—a facility dedicated to discovering how human beings find pleasure. He points out that one of the key functions of the augmented frontal lobe in *Homo sapiens* is to serve as a kind of "experience simulator."

There are, however, several not-so-logical manifestations that stem from our frontal lobe happiness simulators not making correct assessments. People worry much more about an anthrax attack, for example, than the flu. No one has died of anthrax in the last decade, but nearly one million people worldwide have died from the flu each year. But the notion of a malevolent individual poisoning us with anthrax is a man-made risk, so it is more compelling than something like the flu, which seems entirely beyond our control. Similarly, acts of terrorism, such as the 9/11 attack on the World Trade Center, can cause us to make irrational responses. After the attack on 9/11, the United States spent nearly $1 trillion on homeland security over the decade that followed. No one claims that Congress makes sound investments. For such massive security expenses to show an adequate return on investment would require authorities to prevent nearly two thousand terrorist acts per year! Alternatively, if roughly four terrorist acts were prevented every single day of the year, then that would make the spending of such vast sums of money cost effective.

However, we would not be human if we did not recognize that acts of terrorism serve as powerful stimuli to induce irrational and disproportionate responses. While nominally the United States launched its "global war on terrorism" in the wake of the nearly 3,000 lives lost in the

9/11 attack on the World Trade Center and the Pentagon, conservative estimates are that at least 1.5 million individuals have lost their lives in Iraq, Afghanistan, and Pakistan since that U.S. military response began. That would translate roughly into nearly 500 lives lost in the Middle East and Asia for every single life that was taken in the 9/11 attack. In the end, there are times when there is absolutely nothing rational about the way the brain works.

The Stressful Status Quo

The human brain and body have become well suited over millions of years of evolution to handle *acute* stress. Something disturbs our environment and puts us on alert. We run from it, hide from it, or eliminate it. The state of arousal resolves, and the body's physiological and homeostatic mechanisms return to baseline, back to their resting state. In the world of our primitive ancestors, anxious or fearful situations were relatively short-lived scenarios that required reflexive, self-protective instincts to operate unimpeded by cortical intrusion. That was how the system was designed to work.

In today's "jungle"—our world of geopolitical and urban influences— where activity and industry know no circadian limitations, where a "threat" is no longer a saber-toothed tiger but an incessant series of deadlines and shifting world events, there is no rest from stress. It becomes the status quo. It is as if our bodies and brains now find themselves chained to a symbolic saber-toothed tiger 24/7 without any hope of respite. Now there is *chronic* stress—something that our brains and bodies were never prepared to handle. Maintaining this high state of chronic stress and threat readiness induces a debilitating breakdown of the body's defenses. Stress and inflammatory biomarkers damage our blood vessels, leaving us susceptible to heart disease and stroke. They overwhelm our immune systems and deplete their defenses, leaving us susceptible to infections, inflammatory problems, and cancer. We never stand down. Our central nervous systems are being bathed in a never-ending toxic soup of elevated catecholamines (for example, norepinephrine and dopamine) and cortisol.

Bruce McEwen, a neuroscientist at the Rockefeller University, evaluated the effect of stress on neurons and found it inhibits their ability to grow and branch. Chronic stress in rats produces shrinkage of the hippocampi—just as we saw in the brain imaging and autopsy findings in depressed patients with dysregulated hypothalamic-pituitary-adrenal axis. High levels of stress hormone produce what is called dendritic shearing, meaning the rich network of tendrils that allows neurons to interconnect gets cut back—pruned—and reduced in number. Stress, in a sense, reduces our brains to a simpler level of organization and interactivity. In a similar vein, chronic stress also caused stem cells that normally support increased interconnectivity between different parts of the brain to malfunction and reduce the connections formed between the frontal lobes and other areas of the brain. When the brain runs out of avenues of escape from stress, the end result is slow neural self-destruction by degrees. We must stop to consider this: Are we on the verge of creating a world where our brains will be subjected to so much stress they may no longer be able to adapt and leave us prone to failure without extensive adaptative training? *The human brain evolved to be equipped to handle acute stress far better than chronic stress, and the latter can lead to dendritic pruning and diminished brain interconnectivity.*

Now, in the aftermath of the COVID-19 crisis, we *saw what happened when we did* find ourselves in a situation where the threat posed to us by an unknown and unseen enemy felt unrelenting—even existential. Yes, in the early stages of the pandemic, people were told to shelter in place to protect themselves. A kind of communal agoraphobia settled in place, where people became fearful of leaving their houses, of even taking the dog out for a walk. This anxiety level also only seemed to become heightened when people actually did venture outside and were overwhelmed by the sights of everyone wearing face masks and employing social distancing. In the first three or four months of the pandemic, an Amazon delivery person showed up at my door in a full hazmat suit with a full face-shield respirator, the sort of thing you'd expect to see if you were opening the door to a nuclear reactor in Chernobyl. Naturally, he was right to protect himself because he must have had to stop at hundreds of households that could have been potentially infected, but it also added to the subliminal messsage that outside my house lay a hostile, life-threatening world. Therapists reported being inundated with record numbers of clients seeking advice and treatment from chronic anxiety. Levels of alcohol consumption and prescriptions for sleeping pills have skyrocketed since the outbreak.

The COVID-19 crisis seemed tailor-made to induce in us a sustained sense of uncertainty and peril. It was invisible.And initially, untreatable. It was highly contagious and new variants kept cropping up that only seemed to enhance the virus's transmissibility. Finally, new vaccines specifically engineered against the virus began to pour into the marketplace. But, after that, we began to worry because new variants of the virus began to appear that seemed capable of evading the vaccine-induced antibodies circularing in our body. The virus exacted a staggering death toll, especially in the United States, where the per capita death rate was the highest in the world.

The COVID-19 pandemic seemed almost surreal, like something written into a second-rate sci-fi thriller script. However, it has left very real, deep, and lasting scars with a high incidence of PTSD for first responders, frontline health-care personnel, family members, and our children for an entire generation. In its aftermath, we know that the pandemic will leave us with a host of medical issues to deal with for many years after the pandemic has receded.

The COVID crisis exhibited all the stress markers too. It lacked predictability. We had not encountered a global pandemic like this in over a century. The brain shudders at the thought of dangerous scenarios that lack predictability. Remember: the key to training individuals to handling stress was predictability imposed by training. The pandemic was exactly the opposite. Furthermore, we already know the brain weighs negative information more heavily than positive information. In short, we had—and maybe still have—a recipe for a crisis that can create sustained, unremitting stress for months, possibly years.

Our reptilian and mammalian systems usurped control over our neurological responses. Amygdala hijacking is in full force. We were not the cool-headed Captain Sullenberger in the cockpit. Or the steadfast Secret Service agent Tim McCarthy. Sure, there were first responders and health-care workers who were. But the vast majority of individuals were not in control for months after the pandemic started. We panicked. We bought out surplus toilet paper. And we fell prey to denial, paranoia, and tribalism. The death toll here in the United States was a direct measure of how much we lost our way. We did not respond with resolve. In the opening months of the pandemic, the greatest industrial power on earth was sending thousands of first responders, nurses, and doctors—its proclaimed heroes—home to wash out face masks in the wsshing machine with bleach and dry them with a hair dryer because "we just didn't have enough on hand."

No, we cracked and shattered. We did not ward off a threat. We waded into it. At the height of the pandemic, Americans were dying of COVID-19 infection at the rate of more ut two individuals every minute! We developed effective vaccines using mRNA technology in record time, nearly half the country, at one point, was still pondering whether they should get vaccinated. And what did we learn? That even in the face of an existential crisis, our politics and prejudices left us badly divided. Sometimes, even with fatal consequences. And how has this pandemic left us? Do you think that in the next pandemic everyone will respect the signs telling customers not to enter the store without a mask? In the next pandemic, people will not patiently wear their mask aboard a flight or shopping in store. There will be fights and riots against mandates from employers that workers wear their masks and get vaccinated.

One of my colleagues, a physician who was an expert on pulmonology and viral infections, was aggressively booed and then physically chased out of a school board meeting because she explained why it would be premature to do away with facemask restrictions in our school system. She was pursued all the way back to her car, and she was so frightened, she locked the doors, and sped out of the parking lot. Dr. Anthony Fauci, the director of the National Insititute of Allergy and Infectious Disease, and one of the country's most esteemed scientists, received so many death threats for advocating straight, scientific answers for COVID management, that the government had to assign him a security detail for protection.

No. If you thought the COVID-19 pandemic was bad then pray we do not experience another soon. Because our tribal, paranoid, and often plainly ignorant positions have left us unwilling to listen to our nation's leading medical authorities, and willing to lay waste to scientific merit for the sake of political gain. Much of our collective amygdala has been hijacked. We haven't reasoned our way through this crisis and better readied ourselves for the next. We've crippled ourselves instead.

On the Nature of Fear

Brain imaging studies using PET and fMRI during routine intellectual endeavors demonstrate that this is when the frontal lobes exhibit typically maximal activity. This means we can apply the highest levels of brain function in our executive centers to monitor the world and evaluate what we say and do when stress does not interfere with frontal lobe function. Under such circumstances, we have the luxury of thinking before we act. But we must also remember that the neocortex of the frontal lobe is a relatively recent invention on the part of Mother Nature.

As we pointed out earlier, the brain must build on the scaffolding developed over the course of biological evolution. The older, primitive reptilian centers of the brain stem and midbrain were version 1.0 of the original "software" loaded into the central nervous system. Along came the vastly improved, emotionally enhanced mammalian version 2.0 with its limbic lobe add-ons. As primates evolved, they brought along wonderful new features in version 3.0—namely, extensive neocortical capabilities that include new, faster cognitive processing, enhanced memory storage (including emotionally tagged recall), with executive oversight and self-control. As always with the rule of scaffolding, all upgrades were required to be embedded in and compatible with all earlier versions of preexisting software.

The neocortical version of our brain software works well. *But it has a serious bug.* The glitch in the software is that the neocortex will not work when there's too much stress. Because when there is too much stress, the amygdala (the limbic lobe) and the hypothalamus (reptilian brain) become active. As they do, these structures flood the brain with stress-related neurotransmitters: especially dopamine, norepinephrine, and cortisol. As noted, small amounts of these neurotransmitters make our frontal lobes crackle; they are the reason we can think faster, move quicker, and execute better—as predicted by the first, upward-sloping part of the Yerkes-Dodson curve.

However, in overly stressful situations, the frontal lobes become flooded with these chemicals and they become ineffective. They can no longer smoothly integrate with the earlier versions and maintain control over the lower centers of the brain. Quite literally, we stop thinking and just react. Reptilian software version 1.0 takes back over.

S.M.: A Fearless Brain

One way to understand the role that fear and anxiety play in confounding our behaviors is to evaluate a handful of individuals who have a unique genetic disorder that leaves them devoid of a fear response. At first glance, this may sound like someone being given an unusual gift: fearlessness. But it is more like a curse. S.M. is a famous patient in neurology. We know she is an artist, is about fifty years old, and has three children. She has been extensively studied by a stellar team of neuroscientists at the University of Iowa headed up by Antonio Damasio, Ralph Adolphs, and Daniel Tranel. She has a normal IQ and language ability. We know that her childhood seemed quite normal. She distinctly remembers that when she was a child she was afraid of the dark, scared when her brother jumped out from the bushes, and terrified by a snarling dog at the neighbor's. All pretty normal stuff.

As S.M. grew into adulthood, however, she could not recall an episode where she felt fearful. She lived in an inner-city neighborhood where she was once threatened at knifepoint, a second time at gunpoint, and was also the victim of a serious assault. In none of these instances did she ever feel fear. She eventually came to medical attention because she suffered a seizure. The first thing Dr. Damasio noticed while interviewing and examining her was she did not seem to appreciate people's personal space well. She sat uncomfortably close to people. Other symptoms and signs began to appear: a hoarse voice along with small, granular nodules around the eyes. The skin changes were characteristic of a very rare disease called Urbach-Wiethe disease. Only a few hundred cases have ever been found, and eventually all were traced back to a genetic mutation. The disease is also associated with calcium deposits in the brain's blood vessels that eventually damage the medial temporal lobes and destroy the amygdalae. That's right: the engine of fear, flight, and fight.

Neuropsychological testing revealed that S.M. could correctly identify positive emotions (like happiness or laughing) in a series of random-ized photographs but failed to identify photographs that showed negative emotions like fear or disgust. Dr. Damasio, her neurologist, called it a failure to pass "the Doris Day test." "When we showed her a film clip of Doris Day *screaming,* she asked, 'What is she doing?'" The meaning of such images left her mystified. It was well known that the amygdala was activated by strong emotions like fear, but S.M.'s case study also dem-onstrated that the amygdala was used as a reference library for negative emotions. Interestingly, exactly where happy emotions are interpreted is still unknown. In addition, when tested with stimuli that S.M. claimed to "hate," like snakes and spiders, she was able to handle the creatures without hesitation or discomfort.

Post-traumatic Stress Disorder (PTSD)

PTSD usually follows one or more powerfully traumatic experiences. PTSD is often delayed in onset and can occur months (even years) after the trauma is experienced. Typically it is not diagnosed until the patient has suffered with symptoms for one month or longer after the trau-matic incident. It can be associated with trauma due to natural disasters, assaults, warfare, and even traffic accidents. Symptoms include hyper-arousal, avoidance of triggers that remind the individual of the event, disturbing thoughts or illusions of being threatened, and also a high association with suicidal ideation. Individuals who have suffered inter-personal trauma such as sexual or child abuse have the highest likelihood of developing PTSD. Children under the age of ten appear to be less susceptible to developing PTSD.

PTSD is far more prevalent than previously thought. Approximately 3.5 percent of Americans suffer from the disorder. There is a higher inci-dence among those engaged in combat and among women. Further, there is evidence that sexual assault is the leading cause of PTSD among female veterans. It is difficult to predict who will suffer PTSD after trauma, and there appears to be little effectiveness to preemptively treating all those who have experienced the trauma. It is better to aim at early intervention once symptoms develop and the afflicted individual is identified.

PTSD is not a random disorder. About 30 percent of the variance in PTSD is genetic, and there is evidence that some families are more susceptible to its effects. Identical twins have a higher rate of concordant PTSD than fraternal twins. MRI studies have shown that individuals with PTSD have smaller hippocampi and exaggerated startle responses. Studies of blood samples from soldiers suffering with PTSD have white blood cells (which are activated in trauma) with higher numbers of glucocorticoid receptors, meaning that stress could induce an exaggerated response compared with individuals with lower concentrations of receptors. This leads us to a generalized theory about PTSD: genetic and biochemical factors that enhance memory formation and retention will also have a tendency to make the expression and severity of PTSD worse.

During the genocide in Rwanda in 1994, where one in every ten Rwandan citizens were killed, the rate of PTSD among survivors has been heavily researched. Genetic markers indicated that a mutation in the gene that encodes for catechol-O-methyltransferase (COMT) may make individuals *more* susceptible to developing PTSD. Those individuals who had inherited the COMT mutation will have a diminished capacity to inactivate circulating catecholamines—our old friends epinephrine, norepinephrine, and dopamine—all of which will flood the body and brain during trauma. Having higher concentrations of catecholamines may lead to higher and more sustained activation of the amygdala and the rest of the limbic system (see figure 10) during a fight-or-flight situation.

Animal studies looking at the development of PTSD provide some tantalizing clues. For example, beta-blockers (which block catecholamine receptors from being activated by epinephrine, norepinephrine, and dopamine) given *before* the trauma is experienced can dramatically prevent or reduce the incidence and severity of PTSD. Obviously, that's interesting from a research point of view but impractical, because we can't pretreat for a trauma that we can't anticipate.

Similarly, processing during sleep seems to provide a significant step in developing PTSD in ways that are still somewhat mysterious. Administration of compounds that inhibit protein synthesis during sleep also diminish the brain's capacity to exhibit PTSD, suggesting that cellular changes during sleep must be involved in the memory formation surrounding PTSD-triggering events. Longer periods of REM sleep states are associated with decreased risk of developing PTSD in animal studies, suggesting that processing during dream states may help to reduce the severity of PTSD.

As a society, we admire courage. We root for the hero, the protagonist who can keep his head amid the explosions and mayhem around him. In real life, we admire courageous icons like Martin Luther King Jr., Nelson Mandela, or even a fourteen-year-old female education activist, Malala Yousafzai. All showed their ability to remain steadfast to their principles and their passions, even in the face of overwhelming discrimination, domination, and physical violence. The scientific truth is that what we admire in these individuals is their capacity to maintain their cognitive control in the face of subconscious drives of reactivity and fright. In the end, bravery is something that requires inspiration and practice. We need to kindle the fires of our own courage by seeing the true power an individual can exert even under dangerous circumstances. And we need practice. We need to develop the habit of maintaining our focus, of sustaining our cognition, even in situations where the amygdala would wrest away control of our very thoughts. In the end, courage is the ability to see if, as Rudyard Kipling put it, "you can keep your head when all about you are losing theirs."

11

On the Nature of Music

Music is the strongest form of magic.
—MARILYN MANSON

In Africa, I traveled for periods of a week or longer with a handful of Fang tribespeople as my companions through the jungles of Gabon. Individuals would inevitably manufacture musical instruments from natural materials they gathered as we traveled. By nightfall, we were assured of having an impromptu concert and sing-along. The music they shared every evening warded off the real and perceived dangers of the jungle. It generously offered us its collective warmth, solace, and closeness. That is music's great gift: it creates communion among those who share it. From a jug band on a porch in Appalachia to Max Yasgur's field in Bethel, New York, where the Woodstock music festival was held in 1969, sharing and playing music together cements social relationships.

The identity of every culture shines through its music. Archaeologists have demonstrated that human beings played wind, string, and percussion instruments as far back as thirty thousand years ago. That is because music translates into concrete biological advantages. Whether it is a group of workers harvesting crops, or high school students at a pep rally, or soldiers marching off to war, music makes groups cohesive. It also makes them more durable. Scientists at the University of Oxford evaluated groups that sang together. They found that those team members demonstrated elevated endorphin levels. Group members who sing together will exhibit higher pain thresholds and endurance than individuals carrying out a similar activity in silence.

Music as a Universal Language That Unites Us

Music is not, as one neuroscientist put it, "auditory cheesecake." Music *is* magic because of how our brain processes it. Our instinctual appetite for music left even the great Charles Darwin scratching his domed forehead. Music, he wrote, "must be ranked amongst the most mysterious [gifts] with which [humanity] is endowed." It is a powerful force within us: it permits us to understand each other without a translator. We feel the emotion and the meaning of the music far beyond the mere notes or the lyrics. Every one of us has listened to a piece of music that seems to pierce our souls. Music is employed at every solemn ceremony in our lives: from courtship to weddings, from coronations to burials. Music is far more than mere accompaniment. In every culture, music sustains the emotional impact and frames the occasion with solace and solemnity. Music is the soundtrack of our lives.

We see the unifying power of music in our national anthems. In the United States, the national anthem was only officially adopted in 1931 by an act of Congress. Before that time, there was a kind of historical hodgepodge of tunes that were summoned forth on official occasions. Despite its long and indirect route, the national anthem now stirs in many a profound sense of patriotism when it is played. At the conclusion of the Olympics, the games' highlights always include footage of athletes standing on the podium and singing the words to their national anthem while tears stream down their faces. The flag and the anthem bind the athletes to the citizens they represent.

Another place we see the power of music to unify is at rock concerts. Huge crowds (sometimes numbering up to 100,000) can become galvanized by the music they share. It is as if the music has one heartbeat in common—literally. There is good evidence that choirs or choruses that sing together tend to synchronize their heart rates. There appears to be a precise synchronization of cardiac and respiratory rates among the musicians and the listening audience at larger rock concert venues, irrespective of how musically sophisticated the listeners may be. So, music can provide us with an identity and—almost instantly—a sense of inclusion. It not only makes a group feel more cohesive but also synchronizes physiological parameters within the group.

Music and Emotion

Interestingly, there is no one area of the brain devoted to music. Music calls upon substantial amounts of cerebral real estate. It uses more functional areas and engages more brainpower than almost any other human endeavor. Before the advent of modern brain imaging techniques, studying how the brain processed music was quite laborious and fragmentary. We now know from neuroimaging studies that musical sounds reach our ears and are immediately processed by the auditory cortex and the adjacent auditory association area. But once music reaches the auditory association area, it fans out across the brain like a prairie fire. It connects immediately to the frontal cortex (cognition), the limbic lobe (emotion), and the hippocampus (memory).

The memories evoked by music are complex. For professional musicians, musical memories include visualizing the sheet music and even the timing of the hand movements as they visualize themselves playing the music. Similarly, often hearing a few bars of a song can immediately invoke vivid memories of hearing that particular piece of music at a unique time or setting.

Music also serves as an "emotional enhancer." What would a ball game be like if the famous crescendo of the "Charge" theme were not resounding on the organ? How dull would the Batman filmology become if it were not for Danny Elfman's and Hans Zimmer's scores? What's a wedding without Mendelssohn's famous "Wedding March"? A good way to assess the emotional impact of music is to watch famous scenes in some of your favorite movies without the soundtrack. There are websites you can visit that will allow you to watch scenes from *Star Wars* without the illuminating soundtrack of John Williams. Or you can watch Rocky Balboa working out in the streets of Philly without the driving thrust of Bill Conti's score. Or *The Lord of the Rings* absent the music of Howard Shore. It is enough to make you hand your popcorn back in and ask for a ticket refund. Finally, often at funerals, attendees are treated to a selection of the deceased's favorite music. It's like a final glimpse into their personalities. I feel cheated if I attend a funeral and we don't sing "Amazing Grace" together. If they play the tune on the bagpipes, I lose it. It is just too mournful. Yes, it's true. I have been known to slip the piper a twenty to play "Amazing Grace" and "Scotland the Brave." After all, music serves as an emotional enhancer.

Four musical factors help determine our emotional responses to a tune: tempo, loudness, timbre, and tonality. The *tempo* of music directly affects the heart because it tries to sync itself to the music's speed. I always selected baroque classical music with a rhythm of around seventy-two beats per minute while I was performing surgery. It tended to keep my heart rate slow and made me feel calm and in control.

Loudness directly affects both the amygdala and the limbic areas. It commands attention and heightens our emotional state. *Timbre* is also called the tone color or tone quality of music. Timbre is what makes the quality of a clarinet different from an oboe and from a flute, though all three are wind instruments. Timbre is what lends an emotional quality to an instrument through its interactions with our limbic lobe. *Tonality* refers to the arrangement of pitches or chords in a piece of music. In music, we refer to the tonal center as the note that is the center or anchor note of a particular key or chord. The more notes deviate away from that tonal center, the more emotional tension it creates in the listener whose brain is subconsciously looking for resolution. Music that returns quickly back to its tonal center is perceived as relaxing. Many people find Bach's music soothing because it has a high probability of allowing the listener to predict where the upcoming, consonant note will fall. That makes it reassuring and predictable. I knew I had a good reason for liking Bach so much in the operating room. If only surgery could move along as predictably as a Bach fugue.

I often asked patients to make their own soundtracks for falling asleep under anesthesia or awakening from surgery. For the latter, one old Hells Angels biker selected "Sweet Home Alabama" by Lynyrd Skynyrd and told me that he wanted to "wake up like a Polaris missile being shot out of a submarine." I obliged him while he was awakening, but I still played Bach while I was operating on his brain.

Music as Neurological Function

Another function of music is *consonance*. A simple way to think of consonance is when a combination of notes sounds great when they are blended together. Dissonance in the notes makes them sound terrible when played together. This business of sorting out which notes are consonant or dissonant is *not* a conscious one. It turns out that every pitch (that is, every note) activates its own particular group of neurons in our auditory cortex.

Let's take a consonant set of notes: say, you play a middle C note on the piano and a G note (a so-called fifth) together. It sounds lovely to your ear. It turns out the middle C on the piano is a tone with a frequency of 260 hertz (Hz). The middle G note has a frequency of about 390 Hz. Your ear tells you these notes "fit" together nicely. It turns out the ratio of these two frequencies is roughly two to three and our brains like the sound resulting from such a blend.

Now you play the same middle C note, but this time you play the very next note, one step up on the musical scale, the C-sharp note (277 Hz), together. Playing the notes together makes them sound off, almost grating to your ear. They sound *dissonant*. The ratio of their frequencies is seventeen to eighteen. What happens in your brain when you play the C note is that whole populations of neurons tuned to that note (or pitch) are activated and they send their signals on to another set of neurons, called interneurons. When you play the G note, the G note neurons fire, and they send their signals onto the same set of interneurons. Since the notes C and G are consonant, the interneuron fires at a very slow frequency. But when we play the dissonant notes C and C-sharp together, the interneurons fire at a very rapid rate. So, our brains are wired to differentiate consonant notes from dissonant ones.

Many aspects of music appear to be hardwired in the brain; this includes our sense of consonance and dissonance. Babies as young as two months will turn and look toward a sound source when they hear a consonant note being played and will look away from the dissonant sounds. We have other neurons that respond at a higher rate to notes when they are in what we call an ascending contour (that is, notes are arranged in a sequence of higher and higher pitches) and fire at a lower rate when the notes are in a descending contour. So, the very nature of how we process music and detect its qualities is hardwired in our brain in an *unconscious* fashion. Other sets of neurons respond specifically to different rhythms and harmonies. The human central nervous system carefully differentiates and segregates musical functions to specific locales within our cortex. Music can be scrutinized in much the same way as spoken language. As with spoken language, our brain has expectations when listening to where the next word or note should fall. This is one reason we can vividly recall a song or tune from our youth when we hear only three to six notes being played on the radio. In effect, our memory is telling us where, when, and what we expect the next notes to be. That drive in our heads to "name that tune" is about not just conscious recall but also an unconscious drive to hear the notes the way we remember them.

Much of the way that our brain responds to music is determined by our early childhood experiences. Our "musical ear" tunes to cultural norms of music at about ten years old.

Around this time in development, the "tuning" of neurons becomes more or less fixed. Our brain becomes accustomed to certain sounds and kinds of music. This explains why we have cultural and generational expectations in our musical tastes.

If you have ever listened to traditional Chinese music, it can be hard on the Western ear. And it sounds nothing like Beethoven's sweeping, powerful romantic music. In 1906, a Chinese music scholar named Li Shutong tried to figure out how to bring Beethoven to the attention of Chinese audiences. Li had been exposed to Beethoven during his musical studies in Europe, but the Chinese ear was not tuned to the "sound" of Beethoven. Li decided that to make Beethoven more appealing, he would need to tell the story of Beethoven's difficult life. He talked about how Beethoven struggled with poverty and health problems and then, later in life, was plagued with deafness at the height of his creative powers. Li then concluded with the end of Beethoven's life, when he wrote some of his finest pieces but could no longer hear them himself. Something about the hardship and the progress against overwhelming odds resonated with the Chinese public. They fell in love with Beethoven's heroic story. In 1925, at the funeral of Sun Yat-sen, China's first president, a Beethoven symphony was played. Mao Zedong ordered that Beethoven's Ninth Symphony be played in celebration of the tenth anniversary of the Communist Party. In an ironic twist, students leading the famous antigovernment protests in Tiananmen Square in 1989 also played Beethoven's "Ode to Joy" over their jerry-rigged loudspeaker systems. Beethoven became such a revered musician in China that his biography is now mandatory reading for Chinese middle school children. Once considered so hard on the refined Chinese ear, his music is now embraced by one and all. Concerts of Beethoven's music are now instant sellouts in major Chinese cities. In other words, the Chinese ear learned to incorporate the "lexicon" of Beethoven so that it could be more easily understood by the acclimatized auditory cortex of the Chinese listener.

Brain Processing in Professional Musicians Versus Amateurs

The story of what happens to music in the brain becomes more interesting when we look at the brain activity seen on neuroimaging among professional musicians and compare that with the activity seen in brains of amateur musicians as they play the same piece of music. What fMRI reveals is that the professional musician's brain is actually organized differently than an amateur's. There may be a good reason for this: as a group, professional musicians usually start playing music at an earlier age and, hence, practice playing their instruments for much longer than their amateur counterparts.

Professional musicians have far more cortical area devoted to fine motor coordination in the hands than is seen among amateur musicians. The area of cortex dedicated to hand movement in professional musicians can be twice as large as that of amateur musicians. It is not just that the cortex is bigger. It also has a much higher density of neurons. Professional musicians also show far higher levels of cortical activation for a given hand or finger movement on their musical instrument than amateurs. Professional musicians have more cortex, more neurons, and more electrical activity generated when they play music than amateurs playing the same piece. Finally, professional musicians also hear differently than amateurs. Their brains demonstrate faster connections between the auditory association areas and the fine motor areas. The brain of the music professional has remodeled itself to have greater efficiency and faster exchange within the auditory cortex than the amateur's brain. Such unconscious rewiring demonstrates how music can dramatically reshape the brain. So, if we think of these "musical hand regions" as cerebral real estate, then we would see the lots are twice as big, the houses twice as luxurious, holding twice as many residents, who move nearly twice as fast as among the amateur holdings. Tough professional neighborhood to break into.

We now know that the earlier in life musical training begins, the greater its ability to shape the development of these cortical areas. So, we can ask, was it the brain or the prodigy that came first? We know that Mozart composed his Minuet and Trio in G Major at age six. The violinist Yehudi Menuhin had his first solo violin concert accompanied by the San Francisco Symphony by the time he was eight years old. Musical gifts also appear to have a significant genetic component. Professional musicians start playing music at an earlier age than amateur musicians. Their brains develop so that greater cortical areas, more neurons, and faster interconnectivity are devoted to areas relating to listening and playing music compared with their amateur counterparts.

Play, Play, Play

The takeaway message from this neuroscience is that we have highly evolved neural structures to process music. It is not as if only some of us were musical. These studies suggest *all of us are born musicians.* Parents notice that one of the *very* first things that babies respond to in their environment is music. Infants are lulled to sleep by it. Often a child's first use of language is derived from having heard the words of a song. Personally, I have heard—and sung—"Twinkle, Twinkle, Little Star," "The Wheels on the Bus Go Round and Round," and "The Itsy Bitsy Spider" enough times to qualify for some special dispensation to never have to sing them again.

Children often dance in place before they learn to walk. My grandchildren will dance to any music—a Pepsi commercial, a halftime show during a football game (far more engaging with their gyrations and twirls), and even the jingle for the franchise of a local flooring company! There is recent evidence that activity in the musical areas of the brain is closely associated with rhythmic signaling down to our legs. Our ability to orchestrate fluid sequences of movement in rhythmic synchronization to music is no accident. It is a trait also seen in parrots (they especially like the Backstreet Boys, by the way) and sea lions. While, if it makes you feel more exclusionary, humans are the only species that can dance in unison to music. Okay, maybe we're not all as tight as the Rockettes. And while there is a societal meme circulating that straight white men can't dance, that does not appear to be true—at least neurologically.

Music Education

This enormous proclivity for music in our children makes the case that musical education in elementary and middle school should *never* be optional. Music should never be relegated to "an elective." Quite to the contrary, to leave musical education out of the schooling experience would be similar to forgetting to teach a child how to read or write. Leaving a child devoid of the ability to use language would be unthinkable. It should be the same for our second language: music. Every one of us is endowed with musical abilities far beyond anything scientists could have guessed two or three decades ago. Our young children's general lack of sustained and guided exposure to musical education is wasting a set of the talents and abilities that our species honed over hundreds of thousands of years. It is tantamount to leaving our children with a lifelong disability. It is not just their gifts that are wasted but also their ability to enjoy listening to music, to being moved by it—when it sends chills down their spines. The other great pity of the impoverished state of music education in our school systems is that we are losing a great opportunity to stimulate children's brains when they are at their most plastic, when children are most likely to want to learn to play an instrument.

In 2001, the so-called No Child Left Behind Act was passed by the U.S. Congress. It mandated testing in reading, writing, and arithmetic to assess school performance across the country. It also had a chilling and unanticipated consequence: school boards quickly began cutting back on any programming that did not address the core subject areas that would be tested. By 2010, 40 percent of American high schools no longer required that students have access to an arts course of any kind. A 2017 survey showed that 1.3 million elementary-school-age children had no access to musical instruments and musical programming. And, naturally, the vast majority of these children were from poor neighborhoods and school systems. The study, conducted by the National Association of Music Merchants (NAMM), also showed that Black and Latino parents were the two ethnic groups most likely to enroll their children in musical programs if they were offered.

NAMM also demonstrated that children with the greatest level of access to music were less likely to have emotional problems during the course of their school career and did better in other seemingly disparate subject areas like math and science. Children who played music performed better on reading tests; after all, learning two languages, English and music, is better than one. As they matured, children who played music at an early age also demonstrated better communication skills, were better listeners, and showed more empathy than children who were not enrolled in music classes. Children who enrolled in musical education in elementary school were also significantly more likely to graduate from high school. It is ironic that the notion of "no child left behind" would actually mean "but you *can* leave music behind." All of us have natural musical abilities. Within our schools, our children should have access to ample opportunities, encouragement, and instruction to help them find the musician in themselves.

Music: The Original Language

Music is humanity's original language. We see that our brains evolved to have the circuitry to support that language and the responsive plasticity to grow under its influence. It shapes our first steps and then keeps our feet skipping and tapping for a lifetime afterward. The latest revelations of neuroscience also call upon us to make a profound course correction. The primacy of music in our children deserve greater recognition. The universal appeal to relish and play music demand that our government and communities assign the necessary assets and personnel so our schools can teach music and foster the development of our students not just as listeners but also as practitioners and musical artists. And, finally, it is music that magically binds us in our ceremony, our culture, and our play. The great Confucius recognized the primacy of music and wrote in *The Book of Rites* that "music produces the kind of pleasure which human nature cannot do without." It took neuroscience twenty-five hundred years to prove the wisdom of his insight.

12

On the Nature of Memory

Your memory is a monster; *you* forget—*it* doesn't. It simply files
things away. It keeps things for you, or hides things from you—
and summons them to your recall with a will of its own.
You think you have a memory; but it has you!
—JOHN IRVING, *A Prayer for Owen Meany*

The Man Whose Future Disappeared

H.M. was the most famous patient in medicine. Scientists published
more than twelve thousand journal articles about him. In his
heyday—if one can attribute such a notion to a patient's suffering from an illness—more than one hundred different researchers lined
up to carry out tests, experiments, and procedures on him. What made
H.M. unique, what drew researchers to him, was the result of a surgical
mishap that happened on August 25, 1953. But his problems began much
earlier in life.

When he was seven years old, H.M. suffered a mild head injury after
falling off his bicycle. At age ten, he began having mild partial seizures.
By sixteen, his epileptic fits had intensified into full-blown grand mal
seizures with convulsions. As the seizures became more frequent in high
school, he became the object of a great deal of bullying and teasing about
his condition. His treating doctors were caught up in a recent surge in
popular interest in eugenics; they told H.M. to refrain from having sexual
relations because his epilepsy would most likely damage his offspring.

H.M. repaired electrical motors and worked in a factory all of his adult life. By the age of twenty-seven, however, the increasing frequency and severity of his seizures was making it unsafe for him to work on the assembly line, and his employer let him go. H.M. continued to live at home, where his parents looked after him. At about this time, H.M.'s situation was brought to the attention of a flamboyant local neurosurgeon. H.M.'s life was about to take a drastic turn.

The brain surgeon's name was Dr. William Beecher Scoville. He had attended Yale University as an undergraduate and studied at the University of Pennsylvania School of Medicine. He did his residency training at the same institution where I trained five decades later: the Massachusetts General Hospital. As a resident, I can tell you his name was hardly mentioned alongside the other *alumni illuminati* who had trained there. When his name came up, it was almost always with the historical caveat "You know, he was the guy who developed the Scoville retractors"—a kind of surgical instrument still widely used in the operating room. It was only rarely that someone might add, "He was the guy who operated on H.M." That was synonymous with neurosurgical scandal.

Scoville was a descendent of Harriet Beecher Stowe, the author of *Uncle Tom's Cabin*. As a neurosurgeon, he had developed a reputation for performing scores of lobotomies on individuals who suffered from schizophrenia and depression. The procedure was in a vogue among neurosurgeons from the 1930s through the 1950s as a cure-all for almost any mental health disorder.

The Hartford Courant described the Connecticut surgeon as follows:

> Scoville was a fearless—some say reckless—pioneer in developing surgical remedies for a variety of intractable psychological conditions. . . .
> No one who met Scoville forgot him.
>
> While Henry M. is genial, meek and eager to please, Scoville was a sort of James Bond in scrubs who loved fast, expensive cars and motorcycles, a demanding dynamo in the operating room, brilliant at his craft.
>
> "Bill drove fast, lived hard and operated where angels feared to tread," said Dr. David Crombie, former chief of surgery at Hartford Hospital.

Scoville proposed what he would later term "a frankly experimental operation" to H.M. and his parents. It was a procedure he had performed previously in patients with intractable psychosis. It was, in actuality, a totally unproven procedure, and Scoville proposed removing far more tissue than he had ever done in any of his other, earlier procedures. Scoville later admitted to a colleague that he "had a hunch" that the seizure focus was located somewhere in the medial temporal lobes and this was what led him to remove the hippocampus from both sides of his brain.

Although Scoville initially claimed the surgery was successful because the seizure frequency dropped, later it became clear that from the day of surgery H.M. would never make or store another conscious memory in his life. Some have likened it to a real-life version of the movie *Groundhog Day*, but the details are far too poignant to make that comparison work. Until scientists learned of H.M.'s situation—of what the Nobel laureate Eric Kandel later termed an "example of a patient who tragically became an experiment"—the primacy of the hippocampus in making memories was unknown. Nor did surgeons know that they had to maintain at least one hippocampus intact to safeguard memory.

H.M. would never make a new memory again. He could not learn a new word, remember a new face, or sing a new song. If he was introduced to someone new and greeted them, then as soon as he turned away and looked back, that person was utterly unknown to him. He never remembered a new birthday. As he grew older, he could not identify himself in the mirror, because his only memory of himself was when he was twenty-seven years old. When he learned that his father has passed away, he wrote down on a piece of paper, "Dad's gone." He kept the note in his wallet. One attendant remembered H.M. stopped on a footbridge, pulled the piece of paper out, and started to sob. It was like that every time he opened his wallet. He would begin to cry as soon as he learned the terrible news. Then he would forget it. A scientist who worked closely with H.M. noticed one inexplicable detail: after learning of his father's death, H.M. only talked about his father in the past tense.

Researchers found one small break in H.M.'s amnesia. He could learn to draw. At one point, he also sprained his ankle and had to learn how to use a walker. So how did he learn these new skills? Procedural memories are stored not consciously in the hippocampus but subconsciously in the cerebellum, so he could "remember" new motor tasks.

After H.M.'s parents died, a legal battle ensued over who should gain custody of H.M. It was more of a concern because whomever the court appointed as guardian would have control over access and availability to H.M. A researcher named Suzanne Corkin would eventually befriend him and look after him until he died at the age of eighty-two in 2008. Nonetheless, even she was "frank about her strange excitement after his death, when she watched his brain being removed from his skull so that she could inspect it directly at last." Today, there is not a neurosurgeon alive who would remove both hippocampi. *It took H.M.'s tragic story to underscore the importance of an intact hippocampus on at least one side of the brain to maintain adequate memory function.*

His name, by the way, was Henry Molaison.

The Sea Slug: The Secret Lives of Short- and Long-Term Memory

Memory is the holy grail of neuroscience because it wrestles with the existential nature of being as underscored in the equation below:

$$Being = \Sigma \, [Experience \times Memory]$$

Our being is equal to the summation (Σ) of our experiences multiplied by the memories we have derived from those experiences. What makes us different as individuals is what experiences shaped us and how we remember them. One of the funniest episodes that can happen in my family at any large gathering (like Thanksgiving and Christmas) is to ask my brother and me to explain some old family recollection. What always happens is that my brother remembers something entirely different from what I recall. We joke that it looks as if we were raised in two different families, but we weren't. We will discuss later why memories differ so much from one eyewitness to the next. For now, my point is that different versions of the same experiences and the different way we remember them molded my brother and me. Each of us is a unique being produced by a singular and irreproducible set of experiences and memories. No two human beings would be alike even if they started as identical twins.

Very often in neuroscience, the key to solving a riddle is finding the suitable species in which to ask the question. We understand how short-term and long-term memory work at the neuronal level because of the giant sea slug, *Aplysia californica*. Eric Kandel is an Austrian-born neurophysiologist who started working with the sea slug and set off on a lifelong quest to discover how the creature makes memories.

As far as any animal kingdom is concerned, sea creatures have been kind to the neurosciences. The squid, for instance, provided the giant axons that Alan Hodgkin and Andrew Huxley needed to unlock the secrets of the axon potential: the mechanism that all neurons use to create electrical signals. The sea slug was also blessed with large ganglion cells, gigantic neurons visible to the naked eye. The sea slug also has a muscular siphon that it uses to create a water jet to propel itself out of danger.

Kandel and his team could gradually teach the sea slug a *first response*. Whenever a noxious stimulus (usually a very mild shock) was applied to the animal's muscular mantle, it would withdraw its siphon muscle. The researchers repeated the procedure until the slug learned the response

The famous sea slug, also known as *Aplysia californica*. The sea slug emits a dense, colorful cloud of ink when it is disturbed. The sea slug is equipped with a few thousand giant neurons that have made it an ideal animal model to study how sensory information gets processed and converted into short-term and long-term memory.

and would react preemptively. The slug now learned a *second response* (that is, remembered): as soon as it felt a touch on its mantle, it would contract the siphon muscle *before* the noxious stimulus could be delivered. Because it did so, the research team demonstrated that different mechanisms were underlying the two different responses.

The first response occurred when a specific circuit of neurons would repeatedly fire when the siphon was stimulated. Kandel and his colleagues discovered that when neurons repetitively fired, neurochemicals would gradually build up in the synapse so that the neurons would become "trigger happy" and would fire more readily. But as soon as the stimulation stopped, the neurochemicals washed away and the neuron returned to baseline. This is short-term memory. Like when we repeat a phone number over and over until we get to a phone and can start dialing. Stop repeating the number, or have someone interrupt—"Poof!" The telephone number is gone!

But what happens if we repeatedly stimulate the sea slug's mantle? It learns to withdraw the mantle as soon as it is touched. This is the second response, and it's very different from the first: because even if one stops touching the mantle and comes back later and strokes it again, the response is still there. The sea slug has learned and created a long-lasting change in its behavior. This represents long-term memory.

But why does the second type of learning persist while the first response does not? The answer is that in long-term memory genes inside the neurons send a signal down to the ends where cells connect, and grow additional, new synapses. In other words, long-term memory comes about because there are *structural* changes in the cell. They will stay in place, so the memory remains in place. In contrast, short-term memory is simply a *chemical* change—a buildup of neurotransmitters—that get washed away if the circuit is not stimulated. The memory is transitory. And it is long-term memory that holds the key to our being.

Experiential Versus the Remembering Self

Daniel Kahneman is an Israeli American economist who won the Nobel Prize in Economics in 2002 for his work on how our implicit biases affect our notions of well-being and the economic choices we make.

For example, we tend to be more upset when our investment broker calls us and tells us we have lost $10,000 than if he or she calls to reassure us there is still a balance of $100,000 in our account. The sense of a dramatic change for the worse affects our decision making more than the positive reassurance about our investments' stability (that is, the lack of change). Our decision processes are affected by what we remember from earlier events (for example, downturns in the market). This whole process is complicated by the fact that what we remember is not what we truly experienced.

Kahneman points out that what we recall trumps what we experience. For example, we go away on vacation. Our flights are delayed. We are squished into the middle seat, and our seatmate not only is talkative but also has bad breath. We arrive, and our luggage is lost. Our hotel has the air-conditioning knocked out in the middle of a heat wave. But we head to the beach. We go sightseeing. We eat out, and so on. We come home, and our neighbor asks us, "How was your vacation?" We readily answer, "Great!" Why? Because what we recall about the novelty of the vacation, our leisure time, and moments with our family outweighs all the petty frustrations we experienced.

Another typical example would be when we go to a concert hall to hear our favorite symphony. The production is exquisite. As the conductor is winding up to the dramatic finale, a baby in the audience begins wailing. The baby cries through to the end of the piece. As we leave the concert hall, we quip, "That baby ruined the whole performance for me!" The baby actually ruined only the last minute of the concert. Not the whole concert. What the baby *did* ruin was our *recollection* of the performance. What we recollect matters more than what we experienced.

Kahneman elucidated this notion that there are two versions of ourselves. One is our *experiential self,* and the other is our *recollecting or remembering self.* In some ways, we should conceive of ourselves as an experiential engine. The version of ourselves gathers up our sensory perceptions of the world as we travel through it and passes those impressions onto the cognitive self. The latter processes, weighs, and decides what memories from those experiences should be stored. What affects us then is not the experiential self but the remembering self.

Kahneman and his colleague Jason Riis sum it up:

An individual's life could be described—at impractical length—as a string of moments. A common estimate is that each of these moments of psychological present may last up to 3 seconds, suggesting that people experience some 20,000 moments in a waking day, and upwards of 500 million moments in a 70-year life. Each moment can be given a rich multidimensional description. . . . What happens to these moments? The answer is straightforward: with very few exceptions, they simply disappear. The experiencing self that lives each of these moments barely has time to exist. All that survives is whatever memory the remembering self devotes to it.

So, how do our brains store experiences so we can remember them?

Hexagonal Memory

As Kahneman and Riis ask, how can one fit more than half a billion experiences into one brain? Over the last half century, neuroscientists have hypothesized that however our memories are stored, it would have to be in a dense format to encode all those experiences within the limited space available in the hippocampus of the temporal lobe. The key turns out to be one of nature's most beguiling structures: the hexagon.

The hexagon is a universal design throughout nature because it has mathematical magic: it is the one shape that can best fill a given area or volume with equal-sized units and not leave gaps or wasted space behind.

In 2014, the Committee for the Nobel Prize in Physiology or Medicine decided to split the award. It gave one-half of the prize to John O'Keefe from University College London for work he had done in the 1970s. O'Keefe had discovered a new kind of memory cell in the rat's hippocampus. Each of these neurons relayed information about the animal's position in its environment. Each cell represented a unique location in the rat's territory, and O'Keefe labeled them "position" cells. If a rat were moved into a brand-new environment (for example, was placed in a new maze), then each new site in it would have to be mapped out and assigned a new position cell.

The second half of the prize went to a married couple named May-Britt Moser and Edvard Moser. They were neuroscientists who worked at the Norwegian University of Science and Technology in Trondheim. It is not uncommon for a Nobel Prize to be split and awarded to several colleagues to recognize their contributions in a single area of endeavor. In 2014, however, the committee gave one prize for two discoveries about memory that occurred thirty-four years apart. And the hexagon would provide the key to the mystery of how memory is stored.

In 2005, the Mosers found a new type of neuron that helped the rat determine its whereabouts. The cells the Mosers had discovered were in an area called the entorhinal cortex (ERC), which is tightly plastered against the hippocampus. But the brain cells the Mosers found were different from O'Keefe's position cells. These encoded for far more sophisticated information than position neurons. The Mosers' cells worked like a GPS-coordinate system and gave the animal its position relative to a cognitive map or layout encoded in its own brain cells. For this reason, the Mosers labeled this new class of neurons "grid" cells. Another difference between grid and position cells is grid cells are active even if the rat finds itself in unfamiliar territory.

To grasp the difference between position and grid neurons, you need to imagine a mouse walking across stacks of plates in a cupboard. If the mouse freaks you out too much, then imagine it is happening in the cupboard of a distant relative you don't like very much. As the mouse moves about, the position cells in its brain are firing for each unique place in which the mouse finds itself. Plate A and plate B and so on. However, as the mouse moves about, one grid cell will fire whenever the mouse is standing on the edge of any plate, and a separate grid cell will fire when the mouse finds itself at the center of any dish. So, the grid cells help give the mouse its position relative to any plate.

Grid cells have several other intriguing properties. One is *scalability*. The grid cells can shrink the internal cognitive map down to a mouse's modest territory or expand the scale to handle the migrations of humpback whales over thousands of miles. A second property of grid cells is their density. They are tightly packed in the ERC in a hexagonal grid (see figure 11). These hexagons are organized by sensory modality and segregated into groups to handle visual, auditory, tactile, and other types of information. The hexagonal arrangement is believed to provide a tight matrix of synaptic connections that would permit us to weave sensory details into a rich, conglomerate memory.

As a researcher said after he read about the Mosers' discovery of grid cells and the hexagonal architecture, "This changes everything." Why? Because now we can see an architecture that allows for the storage of all types of sensory information that can be commingled and stored in almost limitless combinations. So, for example, that great piece of chocolate cake you had includes a lot of visual imagery ("Hmm, that looks good"), the gustatory experience ("I love the taste of that rich, dark chocolate ganache"), and the auditory ("Hurry up and blow out the candles and someone get him a fork"). Now we are able to visualize the neuronal distribution that could handle all of that data. It is the hexagonal architecture of the entorhinal cortex that permits the dense neuronal organization that provides us with rich, integrated memory function.

Since the discovery of grid cells, myriad types of other subclasses of neurons have also been identified. Some are highly accurate timing cells. Other neurons encode information about the position of our head and limbs. We can also now use fMRI to visualize the activity of these hexagonal matrices in the human brain. We now find that cells in ERC also code for complex emotions and abstract cognition.

Other correlations make sense too once we start to understand how position and grid cells work. For example, one of the prominent symptoms for patients with Alzheimer's disease is to misplace objects of value (for example, "I don't remember where I parked the car"). Commonly, Alzheimer's patients will wander away from home and be unable to recognize landmarks to find their way back. This disorientation would represent a dysfunction of both position cells in the hippocampus and grid cells in the ERC. When the brains of patients who have died with Alzheimer's disease are evaluated at autopsy, researchers see massive neuronal depletion in hexagonal matrices of both the hippocampus and the ERC.

Individuals who carry a gene for a molecule known as apolipoprotein B are at higher risk of developing early-onset Alzheimer's disease. Since there is an assay for this gene, we can identify susceptible individuals long before they develop the first signs of dementia. We find that these apolipoprotein B–positive individuals have marked disorganization of their hexagonal architecture on fMRI compared with a standard control group.

In other words, the propensity of these affected individuals for developing dementia appears to lie in the underlying faults already embedded in the hexagonal architecture within their temporal lobes.

By its nature, memory function has to be complex. The hexagonal matrix allows for an architecture that could explain how an infinite number of experiences (along with our higher cognitive functions) could be encoded to also provide a rich array of associations. The Nobel Prize in Physiology or Medicine in 2014 captured the significance of the two discoveries of position and grid cells. Although widely separated by time and space, the dicovery of these two cell types together provide a breathtaking insight into the fundamental nature and structure of memory.

The Lying Truth

While the underlying architecture of memory may be elegant, it is not immutable. Neuroscientists used to think of memory much like a dead body. Once, it was alive. It was a living, vibrant experience. After that, it was preserved but subject to decay over time. We did not approach memories as if they would ever change. But every time we revisit a memory, resurrect it from the hippocampus, the memory changes. And just the process of accessing that memory means that what we place back into storage is not the same memory we took out. This happens with all memories. It is why stories change and get iconized in the retelling. This is why my brother and I could swear we were raised by different families, in completely different households. We simply do not remember the same experience. This mercurial quality of memory also has serious consequences for those in the criminal justice system.

For centuries, prosecutors have depended on eyewitnesses to produce some of the most emotional and damning evidence in criminal trials. But are eyewitnesses reliable? Do they remember events and individuals accurately? A recent study surveyed 188 judges, prosecutors, law enforcement personnel, and attorneys general and found that as many as ten thousand people in the United States are wrongfully convicted each year. An analysis of wrongful convictions revealed that misidentification by eyewitnesses was the leading factor in more than half the cases. While eyewitness testimony may be the most damning kind of courtroom evidence, it is also some of the least reliable.

Beginning in the 1970s, Elizabeth Loftus began a lifelong investigation into the nature of eyewitness testimony. Her first studies involved reconstructing automobile accidents. She repeatedly showed that eyewitness testimony could be manipulated by subtly feeding information to the witness after the event. For example, volunteer subjects were shown a video of a car accident. Asking participants, "How fast do you think the cars were going when they *smashed* into each other?" elicited guesses at much higher speeds than asking, "How fast do you think the cars were going when they *hit* each other?"

Loftus developed a misinformation paradigm in which she stated that memories were not fixed but rather retrieved, edited, and then re-stored. In addition, within this paradigm of retrieval, memories were susceptible to manipulation and degradation. In a second series of experiments, Loftus sought to address the issue of outright false memories. These memories were especially relevant in criminal cases where witnesses were making accusations based on recalling long-repressed memories.

It would have been unethical for Loftus to see if she could implant the false memory of a crime having been committed. Instead, Loftus and her team decided to implant a specific memory of an event that never happened—a completely false memory: this one involved a particular day at the mall when, as a young child, the test subject had supposedly become lost and separated from his or her family. In nearly one-third of subjects, the research team successfully implanted the false memory to the point that the test subjects were insistent the event had really occurred.

Memories Can Be False but May Still Hold the Key to the Truth

Scientists recently turned to brain imaging to answer the question, can our brains tell the difference between a real and a false memory even if we cannot—at least consciously? Researchers used a dozen normal volunteer subjects and asked them to memorize lists broken down into categories, like animals, trees, or buildings. Later, individual items were presented to the volunteers. Some of the items had appeared on the lists, and some had never been on them. Subjects were then asked to indicate if an item had or had not been on their lists and let investigators know how confident they were about their answers.

Brain imaging studies with fMRI confirmed that when subjects recalled words that had been on the lists and were confident about their answers, there was a dramatic increase in activity detected in the medial temporal lobes, where the hippocampi are located and memory storage occurs. On the other hand, if the subject was wrong or had low confidence in the correctness of his answers, then the medial temporal lobes did not light up. Instead, brain activity shifted over to the prefrontal cortex. Prefrontal cortical activity meant the subjects were second-guessing themselves and asking, "How sure am I about my answer?" When subjects felt sure a memory had been faithfully stored in their hippocampus, no such query in the prefrontal cortex was needed.

Valerie Reyna is a psychologist with a lifelong interest in evaluating the accuracy of people's recollections. Her studies led her to comment once, "Don't let someone's sincerity fool you into believing they are telling you the truth, because people can believe just as deeply and sincerely in a falsehood as they do in the truth—providing they are not intentionally lying." One day, a brain scan may be able to check out the veracity of any witness.

Brain imaging was used in a collaborative study between the University of California, Irvine's Center for the Neurobiology of Learning and Memory and its Law School and the Johns Hopkins Department of Psychological and Brain Sciences to look at true versus false memories. In this study, the researchers showed a set of visual vignettes to the twenty-five participants to remember. Subjects were then scanned and asked about true and false memories. In this case, a false memory was implanted by the research staff. So, for example, a subject might have seen a picture of an individual putting his wallet in his *coat* pocket. Instead, a researcher might implant the false idea the wallet had been put in the *pants* pocket instead. Since the only true memories were ones the test subjects had actually seen for themselves, researchers were able to look on the brain scans while subjects were trying to recall what they had seen. If activation started in the *visual* cortex, it was a true memory because the test subjects had *seen* it with their own eyes. On the other hand, if activity started in the *auditory* cortex, it was a false memory because it was based on something they had *heard* from the staff. Such techniques may be significant in the future in how we query "eyewitnesses."

"Am I Just Getting Old or Is This Something Else?"

When it comes to forgetting, we all want to ask, how can we tell what's just normal aging and what's worse? There is a normal amount of vexing forgetfulness that goes along with the aging process. Why? Because there is some gliosis or scarring that takes place in the hippocampus as we get older. And there is also a gentle decrease in cerebral blood flow. And, finally, the ability of neurons to repair themselves diminishes with age. So, let's put to rest what is normal, non-pathologic forgetfulness and what isn't.

As we get older, we are allowed to forget a name now and then, and we are allowed to substitute our daughter's name for our wife's. But *please* remember that we're not allowed (as I once did) to substitute our dog's name when we are calling one of our children over to taste if the home-made salad dressing is perfect! The "tip of the tongue phenomenon" is normal. We are still okay when we can't remember the name of the actress who played Queen Elizabeth I in *Shakespeare in Love* (Answer: Dame Judi Dench), or who was the quarterback when the Patriots lost to the Chicago Bears in 1986 (Answer: Tony Eason), or the name of the thingamajig that is used for figuring how much to turn that bolt (Answer: the torque wrench). Usually, the word will come to you if you are patient. And, of course, there's "You don't remember where I left my glasses, do you?" That is also normal. And annoying. For both parties involved.

So, when is it dementia? First, my rule of thumb is to ask yourself the following questions:

1. Is memory loss interfering with your ability to function?
2. Is it more than one thing? Are you having problems remembering names and doing math? Do you notice you are having a hard time following the plot in TV programs? These are all more worrisome.
3. Does your family detect a change? Family members know you very well, and they can often pick up on subtle differences that might go unnoticed by the more casual observer.

4. Is it getting measurably worse by the year?

The more intrusive and disabling the memory lapses, the less likely they are to simply be benign age-related loss of memory. If there are reasons you might be concerned about memory loss, bring the matter to your doctor's attention so that formal neuropsychological testing can be carried out. It may or may not be diagnostic, but it can also be useful for establishing a baseline for tracking later cognitive changes. There are some handy cognitive test batteries or inventories of questions you can run through that may help reassure yourself.

Dementia and Alzheimer's Disease

Nothing haunts the generation of baby boomers like the specter of Alzheimer's disease. As Michael Kinsley wrote in a *New Yorker* article on the disease,

> There is a special horror about the prospect of spending your last years shuffling down the perennially unfamiliar corridors of some institution in a demented fog, your diaper hanging loose, being treated like a child by your children, watching TV all day but unable to follow even the most simpleminded propaganda on Fox News or the most facile plot twist of "Downton Abbey." . . . [T]he ultimate boomer race would be the competition to live the longest. . . . [A]s the moment approaches, dying richer will come to seem pointless compared with dying later. . . . I think I underrated the penultimate boomer competition: competitive cognition. The rules are simple: the winner is whoever dies with more of his or her marbles.

The statistics are daunting: if you add up just Alzheimer's, stroke, and Parkinson's disease alone, we have an 18 percent chance of dying with our cognition impaired.* The odds are beginning to feel a little starker, aren't they?

What causes Alzheimer's disease in the first place? It is a chronic, pro-

---

* Even though most people think of Parkinson's as a problem with motor function (for example, tremor), 85 percent of patients affected with the disease will suffer cognitive impairment.

gressive neurodegenerative disease that leads to gradual loss of cognitive memory, then language skills, and advances to a global loss of cerebral function (see figure 12). Eventually, even bodily functions shut down; death usually follows within a decade of the time of diagnosis. From a clinician's point of view, it is one of the most painful diagnoses to make for a patient and his or her family. One day, prevention may be the key, but for now there is virtually nothing that can be done to slow or reverse the disease.

Amyloid is the name of a protein, formally known as amyloid beta, that appears to be the primary etiology of the disease. Amyloid is derived from an *amyloid precursor protein*, the gene for which is encoded on chromosome 21. The vast majority of cases of Alzheimer's disease are sporadic in origin. What we see in this kind of Alzheimer's disease is the genetic marker of an inheritable ε4 allele of the *apolipoprotein E* (APOE). Forty to 80 percent of sporadic Alzheimer's patients possess the APOEε4 allele, and having this allele in your genetic profile increases the risk of developing Alzheimer's disease threefold.

The amyloid forms into plaques. These are focal formations inside the cell that gradually destroy it. A secondary phenomenon in Alzheimer's disease is the development of so-called neurofibrillary tangles. Normally, microtubules are submicroscopic filaments that give cells a scaffolding to lend them shape and also circulate chemicals. In Alzheimer's disease, a protein (called tau) becomes associated with the filaments, making them sticky. The filaments then clump together and form *neurofibrillary tangles*. The amyloid plaques and neurofibrillary tangles slowly choke the cell to death, and neurons then become shrunken, dark, and lifeless ghosts.

We are seeing an alarming rise in the rate of dementia among athletes who have competed in contact sports. Over the course of years, many of rhem have experienced hundreds (and in many cases thousands or even tens of thousands) of blows to the head and, therefore, their brains have been subjected to repeated, concussive traumatic brain injury. What we are finding is that many athletes associated with sports like boxing and football (where blows to the head are common) develop so-called chronic traumatic encephalopathy. Their brain cells express far too much tau and neurofibrillary tangles, and it leads to dementia similar to Alzheimer's disease but with an earlier onset (sometimes even in the late thirties). Perhaps more ominously, studies of young athletes who died of unrelated causes in their late teens but played heavy contact sports like football and hockey in high school already showed evidence of tau protein accumulating in their brains. We must begin to sincerely question the wisdom of letting our young children engage in such contact sports when the repetitive concussive injuries they entail may damn them to suffering from early and irreversible dementia later in adulthood.

A Mind Is a Terrible Thing to Lose

While there may be many causes of dementia, there are few forms of deterioration as cruel. My own stepfather died from Alzheimer's disease. At first, there wasn't that much to notice. He was just forgetful. Real forgetful. He wouldn't remember where he left his sweater. My mom would bring it in from any number of rooms and yell at him, "Stop leaving your stuff all over the house!" and hand him the sweater. Or he would waltz back into the living room with a big grin on his face, beaming over a fresh cup of tea, only to see one already steaming on the coffee table in front of his seat. "Whose cup of tea is that?" he'd ask. "Why yours," I'd answer. "You just made yourself a cup a few minutes ago."

My stepdad began to lose the thread of the news as he would sit there listening. Once Bill Clinton's face came on the TV and was introduced by the newscaster as president and my stepdad turned and asked, "When the hell did he become president? What happened to Reagan?" It got worse. Once he took our dog out for a walk and couldn't find his way home. He started forgetting how to get completely dressed.

Then came the day when he couldn't remember who I was. It was clear that he was trying hard to remember.

"Do you remember who I am?" I asked.

He looked at me and smirked. "Sure. You're a guy I know from Newark, New Jersey," he announced.

"No," I answered. "But we used to live in Manhattan years ago, remember?"

"Yeah, sure," he said. "You just take the George Washington Bridge if you want to get there."

That was it. I reminded him over and over again who I was, but wherever that memory was stored—wherever those hexagons were supposed to be moored—they had broken free and been swept away. The last few times I saw him, it was clear he was growing more and more frightened in a downward spiral of confusion. Just cringing in terror. I swore I would kill myself before I would ever let that happen to me. I just wondered, maybe you could also forget how or when to kill yourself? You would have to do it before you completely disappeared. Otherwise, there would be no one left inside to get it done.

Memory and Being

Memory is the glue that binds our lives and experiences together. Memory provides each of us with the unique and irreproducible imprints lent to us by experience. But as we've seen, memories are biodegradable, subject to the erosive forces of man and nature. One way to visualize how the memory process gives us identity and context in our lives is to examine the artwork of an American painter named William Utermohlen. He was a gifted artist who was diagnosed with early-onset Alzheimer's disease in 1995 and would die from it in 2007.

During the course of his illness, he intermittently created self-portraits. If we assemble these portraits over several years—they stop when Utermohlen became so debilitated he could no longer paint—we see the tragic progression of his vision of himself as he gradually lost his memory and mental faculties. We see his facial attributes (see figure 13) gradually erode until he sees himself as a featureless, almost formless ghost.

This is the haunting embodiment of a life where memory can no longer be sustained. Where recollection seeps away and evaporates, leaving behind the shapeless husk of a life, stripped clean of its identity. To delve into how the process of memory works is to peer into the very heart of what a life means.

13

On the Nature of Time

Truth was the only daughter of time.
—LEONARDO DA VINCI

The Master Editor of Reality

Engineers claimed that television would never get off the ground. The technology was doomed to fail. When the designers started drawing up the prototype TV sets, a seemingly impossible flaw confronted them: images from the screen would arrive a million times sooner than the sound could get there. There was good reason to be concerned. The image from the TV would travel to the viewer's retina at the speed of light; that's 186,000 miles *per second* or about a trillion feet per second. Very fast. The audio signals from the TV would only be able to travel to the ear at the speed of sound. By comparison, this was a poky 768 miles *per hour*—a mere 1,125 feet per second. Engineers argued that this meant the discrepancy would hopelessly ruin the TV transmission because viewers would see the actors moving their lips long before they would ever hear what they were saying. These were fundamental limitations, and they put the whole future of TV in jeopardy.

Engineers had no way of knowing that these concerns would prove trivial. The brain would come to the rescue. The central nervous system reconciled the difference in timing between light and sound quickly. It allowed the emerging television technology to take off, starting with the first regular TV broadcast in 1939 of the opening ceremonies of the World's Fair, presided over by President Franklin D. Roosevelt. He became the first American president to appear on the new medium. It seemed the brain could easily handle a discrepancy in the signals even if it were as large a gap as a hundred milliseconds—a tenth of a second. But how and why? As David Eagleman, a neuroscientist at Stanford University, put it, "If the visual brain wants to get events correct time-wise, it may have only one choice: wait for the slowest information to arrive." The brain had to slow light down so the sound could catch up.

If we take a closer look at the CNS, we see gaps and signal discrepancies everywhere we look. Sensory receptors in our eyes, ears, and skin take a finite amount of time to generate a signal. Then that signal must travel along a nerve, be carried up the spinal cord, and then be relayed to our cortex for conscious processing. All that transmission along meters of nerve fibers takes time. About eighty to a hundred milliseconds. This delay means that whatever events we perceive as happening *now*—this instant—have already occurred in the past.

Think of sensory signals a bit like a miniaturized version of the Pony Express. Riders gallop along the nerve fibers to bring the brain the latest news. And while it all may be new information to the brain, it took time for the riders to get there, which means everything that is relayed to the brain has already happened. The situation is even more complicated because, in effect, we are dealing with riders that are streaming in from all directions, arriving by different routes, and riding at different speeds. A cascade of chaotic sensory messages comes crashing into our brains. We should be seeing a world that is a jumble of signals, reaching our awareness at different times, in different places, and with different latencies. It should be complete pandemonium. But it is not.

The reason is the brain, it turns out, is the master editor of reality. The CNS takes all those chaotic signals, puts the brakes on some of them to slow them down, yanks some backward, shoves others forward, then smooths them out, so they all cross the finish line of conscious perception together, arriving in unison as a team. But we can create some experiments to effectively "catch" the brain in the middle of an "edit." Eagleman and his research team sat a group of volunteers in front of a computer. The subjects would start off pressing a key and then see a light flash on the screen.

Gradually the research team introduced a slight delay—a lag of about two hundred milliseconds—between when the subject pressed the key and when the flash of light came on. The subjects' brains quickly accommodated this delay in a matter of a few trials. After all, the critical relationship to emphasize from the brain's point of view was the cause-and-effect one that existed between pressing the key and the onset of the light. The brain got creative: it reestablished the succession of pressing the key and the light flashing by essentially editing out the delay so it was no longer *apparent to* the subjects.

The researchers then added a new twist: after the brain had made the perceptual *accommodation* of the two-hundred-millisecond delay, they suddenly removed it altogether. The brain had effectively edited out the two-hundred-millisecond delay. That was its sleight of hand. But with the delay now eliminated, this meant that the flash would get bumped up and appear to arrive two hundred milliseconds early! The result was that the test subjects now saw the flash of light as coming on *before* they pressed the key! The central nervous system is constantly editing sensory signals to make the physical world appear more coherent than it is.

In effect, our brains have altered how we perceive the *temporal* relationship of the events in the "flash-lag" experiments. This shifting of time signatures, as it were, demonstrates the brain's profound power to edit the raw, perceptual footage of conflicting signals so that our perception of the world *appears* coherent.

Another, even more sweeping example of our brain's adept editing skills is one we experience every waking minute without even knowing it: blinking.

Where Did the Blink Go?

You and I blink on average thirty times a minute. That blink usually lasts about two hundred milliseconds. Close your eyes for a moment. Then open them back up. Okay, while your eyes were closed, what did you see? Right. It was black. So, if we're blinking thirty times a minute, we should see the world going dark about thirty times every minute. It would begin to look like one of those flickering, old-time silent movies. But we don't. It represents a monumental piece of editing. If we add up all of our "blink time," it becomes sizable. We blink about twenty thousand times a day. That rounds out to about seven million blinks a year. That means we spend about fourteen thousand hours or about two years of our lives blinking! In the dark. But all that flashing, sputtering darkness ends up on the editing floor. Never to make the final cut that gets seen by our conscious awareness. We never even get to see our own blinks.

Our brains seem to have a lot of editorial freedom. But we can ask, what sense of time do our brains truly possess? As I write this, I can hear the ticking of the clock on the wall of my study. That's one way to assess the passage of time. A physicist would say, "Time is a quantifiable unit of measure imposed by a clock." But we have already seen from the flash-lag experiments that this is not how our brains see time. If our brains could answer back, they would say, "No, time is a relative property that emerges from the order in which we perceive sensory information. We edit accordingly." But what happens if something goes wrong with the editor and the editor itself loses track of time?

Losing Track of Time

Patient K.D. was just such an example. She had suffered a minor stroke and seemed to be making a complete recovery at home. Her family noticed she was becoming a little cavalier about her daily schedule. In the past, she had been an early riser but now thought little of staying in bed past 11:00 a.m. and then having pancakes for breakfast at 1:30 p.m. She had always been an avid baker, but now all her recipes were turning into disasters. Loaves of bread that were supposed to rise for hours were suddenly given only ten or fifteen minutes. Butter that was supposed to come to room temperature for a few minutes was left out overnight. If she forgot to set a kitchen timer, things inevitably got burned in the oven. The smoke detector went off almost every day.

K.D. had always been an avid pianist who often played for an hour or two at a stretch. Now her practice sessions could be as short as ten minutes and, on a few occasions, had lasted as long as five hours. Once, her husband showed her a clock and pointed out she had been playing for more than half the day. She said that could not possibly be true. Something was wrong with the clock. She once took the dog for a walk and never came home. She was found more than six hours later and had traveled more than eight miles from home when the police found her.

The medical term for K.D.'s condition is "dyschronometria." A simple enough solution would be to wear a watch, right? But K.D. was becoming increasingly suspicious of watches and clocks because, in her eyes, they seemed to be increasingly broken. K.D.'s stroke was mild and had occurred in her cerebellum, a relatively silent and ancient part of the reptilian brain. It is usually involved with integrating the more or less automatic mechanisms of muscle coordination. The timing cells in the cerebellum make sure muscles fire in the proper sequences, so you are not, for example, releasing the baseball before you've pitched it. One hypothesis about why dyschronometria can occur with cerebellar injuries is that there may have been damage to the timing mechanisms in the cerebellum that are related to muscle coordination.

An Onion of Timekeepers

When it comes to time, the brain is an onion with layer upon layer of timekeeping systems. We have a clock that sets the day's circadian rhythm. Think of it as the brain's version of Big Ben. We'll be discussing it shortly. Besides this one simple, big clock, there are smaller clocks throughout the brain. In addition to the cerebellar ones that affected K.D.'s sense of time, accurate timekeeping cells are embedded in the hippocampus and parahippocampus. Some of these serve as the "time stamps" for our memories as we store them. So, the recollections we made of last year's Christmas are not mistaken with, say, the memory of a Christmas from our childhood. We don't confuse them because the time-stamp function in the temporal lobes gives us a good sense of when they occurred in our lives.

Beneath the cortex, distinct gray matter clusters, called nuclei, lie close to the brain's center. Each one is a dense collection of neurons. Neurologists call them collectively the striatum. Timing cells of various types populate the entire striatum as well as the hippocampus and parahippocampus. These are the brain's stopwatches.

These timing cells can cover everything from an interval lasting only a few milliseconds to a few minutes in length. For example, when we hear a sound, we immediately compare how quickly it arrives at our right ear versus our left. That difference can be as little as ten microseconds, but our brains can analyze it to tell us from where the sound is coming.

The real wizards of sound discrimination and time, however, are professional musicians. Horologists (time researchers) love to test them—they especially love drummers—to see how small a difference in timing they can detect. Researchers routinely find some professional musicians can differentiate intervals as short as a fraction of a millisecond. Expert percussionists have acquired a finely honed ability to discriminate the timing of events; this is especially true when it is associated with tapping, drumming, or even dancing. This link between movement and timing is something we hinted at earlier concerning cerebellar timing. But it is also why you see so many people tapping their fingers or toes or rocking back and forth as they listen to music. The brain loves to marry movement and timing. Many think the drummer with the best timing was none other than Ringo Starr of the Beatles. Professional drummers point to his achievements like the song "Good Morning, Good Morning," on the 1967 *Sgt. Pepper's Lonely Hearts Club Band* album, where the beat count is changing with almost every measure in the song and Ringo's timing is flawless.

Recent research has demonstrated that the ability to parse time expertly is dependent upon our dopaminergic function. There is a great deal of indirect evidence for this. For example, dyschronometria is associated with impaired dopamine function in dyslexic subjects. At first glance, this might seem a bizarre connection: reading and time. But remember: the brain does not work the way we think it should. It plays by its own rules. Dyslexia is a family of disorders, but one aspect they share is impaired dopamine function and disordered timekeeping. Dyslexic individuals struggle with the amount of time their brains allow them to focus on what they're reading. If the timing and duration of one's concentration are compromised, then the ability to process and integrate visual signals is impaired. Recent research shows that dopamine directly affects the genes that control clock function in neurons. Dopamine pharmacotherapy can help improve dyslexia by promoting more sustained visual processing and memory. Altered dopamine metabolism also occurs in Parkinson's disease and attention deficit hyperactivity disorder; both conditions are linked to lower concentrations of the neurotransmitter in the striatum. In laboratory testing, we see an impaired ability to discriminate between equal and unequal time intervals. Again, when pharmacological therapy is employed to improve the availability of dopamine, these deficits can disappear completely. *Dopamine appears to be intimately involved in the control of neuronal timekeeping functions.*

"Hearing Lips and Seeing Voices"

Earlier, we discussed how the brain creates coherence in our world by synchronizing asynchronous sensory information. In 1976, British psychologist Harry McGurk published an article titled "Hearing Lips and Seeing Voices." He described what happens when we look at a video of an actor saying *one* word paired with the audio track of him pronouncing a second word. The result of this mismatch is that we hear a third sound, entirely different from the words used in either the video or the audio track. This is now called the McGurk effect. It is mainly used to dissect how the brain uses visual and auditory information as part of spoken language processing and is especially relevant in reading lips.

However, in 2013, the McGurk test was used at Queen Square Hospital in London to evaluate a unique patient, known by the initials P.H. He was suffering from a syndrome that doctors had never seen before. The patient was a sixty-seven-year-old retired pilot who had noticed that a few weeks earlier he had begun to hear the audio signal from the television *before* he saw the actors moving their lips. It was the exact opposite of what had worried the first designers of TV. Furthermore, P.H. soon noticed he had a similar issue when listening to conversations with other people.

Using the McGurk effect, neurologists pinpointed that P.H. had a 210-millisecond disparity between first hearing and then seeing. Probably even more disturbing was that he soon began to hear his own voice before he could feel his mouth, lips, and tongue move to make the sound. MRI studies showed that he had suffered a stroke in one of the critical pathways in the brain stem that link auditory and visual information before relaying it to the cortex.

Gustav Fechner

In the nineteenth century, scientists began teasing out the link between physical properties and perceptual ones, a discipline known today as psychophysics. Gustav Fechner (1801–1887) was a German scientist and is considered the father of the field. He first demonstrated that spinning a white disk with black lines and shapes on the surface made observers see colors emerge on the disk. The colors, however, are an illusion (now called the Fechner color effect). It is not an optical illusion so much as a visual one, meaning it originates from the brain's processing abilities. The detection of color is a result of our brain's old problem with asynchronous signals. It arises from the visual cortex detecting minute disparities between the incoming transmissions from different color receptors in the retina as they are stimulated at varying rates by the rotating pattern. Occasionally, the same effect can occur when looking at the blades of a rotating fan or a propeller. Fechner's discovery established that *the timing of our perception alters the apparent physical properties of the world around us.*

Time Is Slippery

Time is a slippery business—even for mainstream scientists. Physicists and philosophers alike have found it is nearly impossible to know what time *really* is. For centuries, we seemed to have the matter reasonably well in hand. In 1721, a meridian was established at the Royal Observatory in Greenwich, England. It was an imaginary line spanning the globe from the North Pole to the South. It passed through the very heart of the observatory and was eventually used to establish Greenwich mean time (GMT). GMT was used to set clocks all over the world and became known as the prime meridian. Today it is preserved by a spherical sculpture balancing on a stainless-steel strip in the observatory's courtyard. And it still might be in use today if scientists had not discovered that the planet is ellipsoid and not a sphere due to distortion from Earth's rotation. This meant that the time in Greenwich would be ever so slightly different from time, say, on the equator, where the planet bulges outward. All of this ellipsoid-bulging business led scientists to decide in 1972 that

GMT should be replaced by an atomic clock based on the nine-billion-plus oscillations per second in a cesium 133 atom that is kept at a temperature of 0 degrees Kelvin (minus 459.67 degrees Fahrenheit—more or less). One of these cesium clocks takes up an entire floor of a large building and looks more like a nuclear reactor than a timepiece. And a lot less sexy than GMT. Then scientists discovered that minor variations in Earth's rotation also introduced minute discrepancies in a single atomic clock, so we currently maintain no fewer than four hundred of these frigid behemoths all over the globe so their results can be averaged together. So much for going down the rabbit hole of finding "a quantifiably finite unit of time."

What we do know is that about ten billion years after the big bang, life emerged on our planet and eventually gave rise to our species, *Homo sapiens*. We know that as primates we evolved over nearly four billion years to be faithfully entrained to the movement of our planet. It has been Earth's rotation that has always set our circadian rhythm. And it is the most fundamental formulation of time our bodies can read and respond to. It is noteworthy that it is only in the last century that we have posed significant challenges to the meaning of circadian rhythm.

The Brain's Circadian Rhythm

The brain's Big Ben is an area known as the suprachiasmatic nucleus (SCN). It gets its name because it sits above the optic chiasm, where the optic nerves from both eyes intermingle. Here, the SCN can record the light-dark cycles in our environment. And it binds human life inextricably to the processes of daytime and nighttime. The SCN, in turn, sends out signals to coordinate the time signatures of cells in the rest of our bodies.

We eradicate the diurnal/nocturnal rhythm if we destroy the SCN in an experimental animal. If we transplant an intact SCN back into the animal, then its normal rhythm is restored. When we dissect individual neurons in the SCN and grow them in cell culture, we see each cell is firing at its own rhythm in isolation. But as the cells grow and spread out and come into contact with each other, the SCN cells synchronize together. In the SCN, we have a few thousand neurons firing together in a synchronized rhythm to generate our daily pattern. So, each cell in the SCN may "tick" to its own beat but "tocks" in unison to create a perfect, twenty-four-hour-long cycle. Well, almost perfect. The way the brain maintains tight synchronization in this multicellular, symphonic timepiece is to make sure the cells get recalibrated each day with the light signals—called zeitgebers (from the German for "time giver")—traveling in the optic nerves. *The suprachiasmatic nucleus establishes the brain's diurnal rhythm.*

Professor Joseph Takahashi, a leading investigator in biological time, has called this molecular mechanism the way "a day is created within the cell." This circadian apparatus has been vital to the evolution of life on our planet, and it dominates the behaviors of almost every single life-form on it. From bacteria to plants, from dinosaurs to humans, all life derives its rhythmicity from Earth's rotation around the sun and the light it sheds on our planet's surface. The light changes the expression of genes within the cell's nucleus. It is what tells a fungus to grow in the dark or a flower to open in daylight. The rhythm of life had been dictated for millennia by the twenty-four-hour rotation of Earth on its axis. For humans, this meant our activities were constrained by our ability to have enough ambient light by which to work. It is one of the harsh lessons that Nature reminds us of every time we venture into the wild. We must make camp, pitch our tent, get our belongings nestled into it, and get dinner going while there is still "daylight to burn," as cowboys are fond of saying. At night, we settle in around the fire, swap stories, and turn in to sleep until daybreak. We have to wait for sunrise to have enough light to see and get back to work. That is until everything changed in the year 1879.

Incandescence: Artificial Illumination at Night

In 1879, on New Year's Eve, a special railroad train had left New York City full of society's most influential couples and celebrities. Three and a half hours later, it pulled in to Menlo Park, New Jersey. Just as the New Year was about to strike midnight, a string of incandescent bulbs erected along the town's main street suddenly came to life and ushered the guests toward the front entrance of the factory Thomas Edison had created at the end of the road. When they walked into the main reception hall, the guests were dazzled as dozens upon dozens of incandescent bulbs twinkled magically.

This was Edison's first public demonstration of his lightbulb. The *New York Herald* reporter announced that Edison had manufactured an incandescent bulb capable of creating a light "like the mellow sunset of an Italian autumn." For the next ten days, capacity crowds flooded into Menlo Park to witness the great phenomenon for themselves. "The Wizard of Menlo Park" had succeeded in turning "night into day." And Edison forever altered the rhythm by which we live and work.

The Problem with Lightbulbs

Artificial light allowed industries and offices to introduce night shifts to enhance productivity and the availability of services. Lights were introduced into cities, making them safer and changing how businesses catered to patrons at night. The effects would change the face of civilization itself. But it also had the power to derail the powerful mechanisms humans had evolved to synchronize the SCN to dark-light cycles. Workers who

The continental United States, as seen from the NOAA satellite showing the cities illuminated by artificial light at night. Nighttime illumination would go head-to-head with the sun to see what would have dominion over the diurnal cycle established by the SCN.

chronically staff the graveyard shift pay a price for working at night: they have a higher rate of coronary artery disease than age-matched controls on the day shift. Similarly, night shift work is associated with an increase in the incidence of cancer, diabetes, and obesity. Individuals who work the night shift also exhibit a higher rate of developing mood disorders, suffering from depression, and show diminished cognitive function compared to workers on the daylight shifts.

The Brain's "Big Ben": Wake-Up Call

What sets the clock for the bossy master timekeeper is no trivial matter. The SCN orchestrates an enormous variety of hormonal and metabolic bodily functions. They include the secretion of melatonin and the body's core temperature. Our temperature drops from 98.6 degrees Fahrenheit to about 1.5–2.0 degrees lower. It will reach its nadir somewhere between 4:00 and 6:00 a.m. This drop in temperature, in turn, triggers the activation of our sugar stores and revs up the immune system. The earlier in the morning your temperature falls, the more of a morning person (an early bird) you are likely to be. The later it happens in the morning, the more of an evening person (a night owl) you are likely to be.

Circadian Rhythm: Desynchronosis (a.k.a. Jet Lag)

Even with the tug of artificial light, our bond to Earth's twenty-four-hour rotation is palpable. The farther we go east or west, the greater the challenge to our preexisting daylight cycle when we take a trip. And it takes time for the SCN to reestablish a new circadian rhythm. A rough rule to calculate the days required by the SCN to adjust is:

Days for SCN to recover = Number of time zones crossed ÷ 2

So, for example, if I am traveling from Tucson, Arizona, to Paris, France, There will be a eight-hour time difference, so the calculation should look like this:

Days for SCN to recover = 8 ÷ 2 = 4 days

So, I should set aside four days (it is Paris, after all) just to recover and allow my SCN to reestablish its whole circadian rhythm. The severity of jet lag can also be reduced by self-administering melatonin on the day of departure to the target bedtime at your future destination.

Concerns about jet lag go far beyond counting how many more museums or bus tours one can take in. Jet lag (medically known as desynchronosis) can pose serious medical problems because it represents significant physiological stress on the body. Heart disease, including infarction and congestive heart failure, has been aggravated by jet lag. An Israeli study demonstrated that the more time zones a person crosses, the more likely they would be to experience a worsening of depressive and psychotic symptoms. A separate Australian study documented that there was an increased rate of suicide in association with even the relatively small shifts imposed by daylight savings time.

Perpetual Night: The Uncoupling of Circadian Rhythm

To see how vital daylight is to the synchronization of the SCN, we need to go underground—down into the dark, alien world of subterranean caves. Living deep underground is much like living in outer space. This juxtaposition of inner and outer space captured the imagination of a young Frenchman named Michel Siffre. He knew astronauts would have to endure hardships in deep space like isolation, light deprivation, and cold. Living in an underground cave could simulate all of these obstacles.

In 1962, Siffre decided to carry out a singular experiment to prove his point. He descended to a cave more than 375 feet underground. Temperatures in the cave were well below freezing. He camped in a tent with only one-way telephone contact with the surface. He had no clocks. No calendars. The only light he saw was what his batteries could provide through a lantern and a flashlight.

Researchers located in a command center on the surface were careful to ensure that no one inadvertently communicated any external cues about time of day to Siffre. The plan was for Siffre to check in every day with the surface team and to remain underground for thirty straight days. Siffre kept a journal of his activities, and he was shocked when his colleagues showed up at his cave to retrieve him after he had recorded only twenty-eight days had passed. Initially, he was angry the research team had violated the protocol. In the isolation of his cave, however, Siffre's "days" had grown much longer than twenty-four hours and, in some cases, were more than forty-eight hours long. Siffre was shocked when his team informed him that he had been underground for sixty-three days! This freewheeling circadian rhythm will be one of the major concerns for any medical officer posted on a mission to Mars.

Siffre's experiment demonstrated that the brain would cycle crazily without the benefit of light. His brain had difficulty maintaining sleep-wake cycles. His dream states were profoundly altered, and he began to suffer from acute depression and vivid hallucinations. Siffre's experiments made it clear that there was far more to face in deep space than just zero gravity.

Deep Space: Where There Is No Diurnal Rhythm

Long before human beings were traveling in space, they had to contend with life aboard submarines. For decades, the navy has been using eighteen-hour work schedules on submarines, which allowed the crew six hours of sleep each night for every twelve hours of work. Unfortunately, it still leaves sailors jet-lagged six hours a day and suffering with problems of chronically inadequate circadian rhythmicity.

Space travel may represent an even more significant challenge in overcoming jet lag. Currently, astronauts orbiting Earth are faced with alternating sunrise and sunsets every ninety minutes. This results in profound disruption of circadian rhythm and leaves astronauts with serious sleep deficits, associated with decrements in alertness, reaction time, and cognitive problem-solving abilities. Low-intensity lighting has been employed on the International Space Station, but unfortunately it is below the threshold that entrains adequate circadian rhythmicity.

Laura Barger, a sleep researcher at Harvard Medical School, evaluated what problems with circadian cycling astronauts would face in adjusting to a 24.65-hour-long day on the planet Mars. That does not seem like much. But remember, the human brain has been shaped over billions of years to a 24-hour-long day. Even when we suffer from jet lag here on Earth, our brains still reset themselves back to a 24-hour day. Not 25

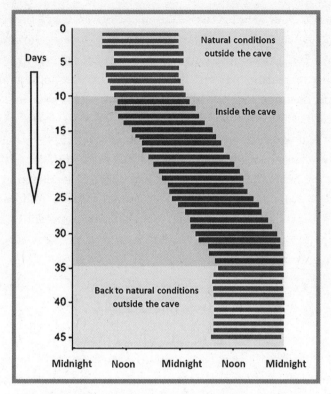

An "actogram" that charts Siffre's activity from days 0–45 during his cave-dwelling experiments. The dark horizontal bars indicate when Siffre slept. In the upper part of the panel, Siffre was outside the cave, and his activity was synchronized by natural, external cues to a twenty-four-hour-long cycle. In the middle panel, Siffre was underground in a cave, completely removed from temporal cues, and his daily cycle began to drift and lengthen in its periodicity (and shifts to the right). In the lower part of the panel, natural, external temporal cues are restored, and Siffre's activity quickly re-synchronizes to a twenty-four-hour-long cycle. Nearly half of blind individuals have also been found to suffer from the same de-synchronization that cave dwellers do.

hours. Barger wanted to see if astronauts could adjust. The answer was no. The astronauts' brains stubbornly refuse to adapt to it. "[The astronauts'] circadian rhythms aren't able to adjust," said Barger. "They have problems sleeping, and everyone walks around with a bloodless pallor." Her research demonstrated that it would take astronauts more than a month to adjust to the additional thirty-nine minutes added to each

Martian day. Today, the European Space Agency maintains a facility in a deep cavern in Sardinia where astronauts undergo psychological training to ensure they can handle the circadian challenges that space travel presents. The introduction of artificial light, plane travel, and space flight have all created profound physiological problems derived from disruption of diurnal rhythm.

From the perspective of outer space, it is easy to understand how we are bound by the biological imperative of our planet's twenty-four-hour-long day and a daylight/nighttime cycle. Breaking away from these will most likely prove near impossible. And while a prolonged day may be difficult to shift in the brain, we routinely see hours and minutes that seem to fluctuate more with our age, mood, and level of engagement.

Scared into Slow Motion

All of us have had the experience of being in a car accident or some other trauma where we see the whole event unfold before our eyes as if in slow motion. I once had a horse kick me quite forcefully in the leg. To me, it all seemed to happen in slow motion—even though horses kick with lightning speed. As I saw the horse's leg flex, I could see I wasn't going to be able to get out of the way. And it was clear to me the kick was going to do a lot of damage. That assessment was accurate; the horse broke my leg in three places. But as the horse's leg came out toward me, I remember I could see every vivid detail of her hoof, right down to the size and shape of the nail holes holding her horseshoes in place. After she struck me, I collapsed on the barn floor and dragged myself to the house to get help.

But we can ask, why do such events unfold in slow motion? Well, first, we're usually scared out of our wits. As I was. The fear fully activated my fight-or-flight response. There was undoubtedly maximal amygdala stimulation. A flood of dopamine was soaking every memory that I was making. David Eagleman carried out a series of spectacular experiments in which volunteers (I would not call them normal) fell backward from the top of a fifteen-story tower into a net 150 feet below. The protocol was designed to instill some of the fear that seems to be integral to such experiences. Careful experiments with precision timing devices proved that absolute, physical time did not change as the volunteers fell from the top of the tower. But their subjective sense of time did. Their sense of time seemed to become prolonged throughout the course of their fall: they felt as if the fall lasted forever and everything were in slow motion.

About 80 percent of us have experienced a traumatic event that seemed to unfold in slow motion. A factor common to almost all of these experiences is a hefty dose of adrenaline. It seems to be a prerequisite for helping our brains to feel as if time were slowing down. What is clear is that we remember such events more vividly. And that is believed to be the solution to the conundrum: when we are emotionally aroused—but particularly when we are in danger—we remember more details because we are making memories faster than usual. It is akin to shooting film at a higher speed. We are effectively making more memories per second the way one would expose more frames of film per second. The rate of memory formation is the speed of life. When we make more memories per second, life slows down.

We discussed PTSD in chapter 10. It is important to point out that activation of the amygdala and fight-or-flight response along with all of its hyper-adrenergic drive set the stage for amygdala hijacking that bypasses frontal conscious processing. But it also sets the stage for a very detailed memory to be stored in scenarios of PTSD. So, the memory trace left in the hippocampus is far richer in detail and far less processed than the average memory. And, unfortunately, the better your memory works, the more likely you are to experience PTSD. Finally, fear affects the speed with which memories are created and stored and alters the subjective perception of time passing more quickly or slowly. But there are other factors to also consider.

Changes in Time Perception

Our sense of time doesn't just change when we go underground. Or when we are scared. Our mood can affect it. And as we age, the pace of time can accelerate. And there are rare moments when our brains also bring time to a standstill.

Does Time Fly When We're Having Fun?

We have all heard the expression "Time flies when you are having fun." Or, the converse of that maxim, "A watched pot never boils." The pot *will* come to a boil as dictated by the laws of thermodynamics, but because we are, presumably, doing nothing but staring at it, the water *seems* to take much longer. I suppose the more up-to-date, modern-day equivalent is computer software never seems to finish downloading while you are staring at your screen. But that begs the bigger question: Does our perception of time change when we are having fun and slow down when we are bored?

Two psychologists, Philip Gable and Bryan Poole from the University of Alabama, took to the lab to see if they could test the "fun-time-flying" hypothesis. Subjects were asked to look at three sets of images. The first set was emotionally neutral (for example, a simple shape). The second was of low interest (say, a flower). And the third set was considered of high interest (for example, an ice cream cone).

Subjects were asked to guess how long the images were on display. Neutral or lower-interest photos were judged as being displayed for longer intervals than high-interest images. Also, when test subjects were hungry, they felt the dessert images were being shown for even shorter periods. So, we can say time flies, at least when we are hungry.

Does Time Accelerate as We Get Older?

A second phenomenon we experience is that time seems to go faster as we age.

We can all recall the experience of being released for summer recess when we were in elementary school. The summer seemed eternal, as if it might last forever. But as we age, summer becomes briefer, almost fragile, scarcely acknowledged before it fades.

A French philosopher named Paul Janet (1823–1899) suggested that we measure time in proportion to how long we have been alive. This has been labeled the log-time hypothesis. It states that if I have been alive for only ten years, then a one-year period would represent one-tenth of my entire life span. So, I would perceive that year as a relatively long time. By contrast, if I were to reach one hundred years old, then that same year

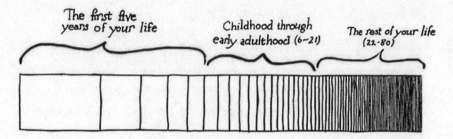

A diagram demonstrating the so-called log-time hypothesis—namely, that while we age at a uniform rate, the passage of time seems to accelerate exponentially.

would reflect only one-hundredth of my whole life, so it is proportionally ten times shorter for me at age one hundred than at age ten. According to Janet's hypothesis, since we age at a uniform rate, the passage of time appears to accelerate logarithmically as we get older.

Passage of Time and Novelty

David Eagleman, who spends a lot of time thinking about time, tried to tackle the same issue. He put forth the hypothesis that when we are younger, everything is new and novel. We, therefore, dedicate a lot of "memory space" to those details. Like our first kiss. Or our first car. Then, as we accumulate experiences, we devote less memory space to new occurrences because there are fewer of them. There is less that is new and novel to our perception. This goes along thematically with Eagleman's earlier "tower terror" experiments that demonstrated time slows down when we feel threatened. Hence, we end up laying down memories at a higher rate of speed. So, the converse could be true. We could say to ourselves, "I've seen this all before. It's uninteresting to me." So, time moves at a crawl because we are laying down memory at a crawl.

Passage of Time and Temperature

Although the pace at which time unfolds may appear to speed up as we age, there does not appear to be any difference in how relatively short periods of time—an hour, a week, a month—appeared to test subjects ranging in age from teenagers to ninety-year-olds. Hudson Hoagland, a psychologist, was tending to his wife, who was ill with a fever. He noticed that his wife seemed to complain about his being gone from her bedside more when her temperature was elevated. The higher her body temperature, the more time seemed to slow down. He repeated this experiment by having test subjects put their heads inside a heated helmet. He showed that subjects' sense of time could become distorted by as much as 20 percent with elevated temperatures.

Passage of Time and Saccadic Eye Movements

Another researcher has tied time literally to the beat of our eye movements, called saccades. Saccades are minute jerking movements of the eyes that bring them back to a fixation point several times each second. The speed of saccadic eye movements gradually increases as we age. In that sense, we would effectively be processing fewer images per second as we get older. That returns us to that notion *that less information processed per second* effectively gives us the sense of the passage of time accelerating. So, we have seen that numerous factors can affect our perception of time; they include fear, age, novelty, fun, boredom, saccadic eye movements, and temperature.

Can We Make Time Stand Still?

How and when does the brain make time stand still? There seem to be two diametrically opposed answers to this question. The first is flow theory. And the second is sensory deprivation. As we'll see, they are almost polar opposites of each other. And a third aspect that may straddle the two is mindful meditation.

Flow States

Flow theory was put forth by a thoughtful psychologist with the nearly unpronounceable name of Mihaly Csikszentmihalyi (pronounced *muh-hei-lee chick-sent-mee-hai-lee*). Csikszentmihalyi wanted to understand *flow states*. Flow states occur in those wonderful moments when we become so wrapped up and absorbed in what we are doing that time seems to stand still. And they are among the happiest moments of our lives. These are hardly moments of relaxation but rather instances of intense focus.

A flow state was just what I experienced performing surgery. I used to joke with my fellow surgeons that performing an operation well made me feel high, as if I were floating three feet off the ground. I often have the same sort of feeling when I am sailing or skiing. Csikszentmihalyi found that the phenomenon of flow states was the same for a brain surgeon in Tucson as it was for a Sherpa in Nepal. Eight characteristics help define a hyper-focused flow state:

1. We experience intense concentration in the present moment.
2. Our actions and awareness become one.
3. We lose our sense of reflective self-consciousness.
4. We sense we have masterful control of the situation.
5. Time seems to stand still.
6. The activity is autotelic, meaning it is its own reward. We do it for the sake of doing it.
7. We feel that we can stretch our skills but still succeed.
8. We lose a sense of all our other needs.

I found flow states very easy to identify when I was performing surgery. First, since brain surgeries are usually very long surgeries—they can stretch on for hours—my flow states during operations would be very sustained. In such a condition, I could not feel my feet ache. I became unaware of any pain in my back from standing for so long. I would be working away and always be shocked to discover a new team had scrubbed in after shift change. I was oblivious to how much time had passed. I didn't feel my bladder. I didn't get hungry. And, as weird as it may sound, I felt utterly absorbed by the brain I was working on. Every detail of the tissue mattered to me, and I was enthralled by it.

Flow states emerge out of total absorption in the task at hand. Since the brain can handle only about 200 bits per second of information, as individuals become more focused on what they're doing, the brain edits out more and more sensory information to keep that bandwidth manageable. It begins by editing out sensory input (for example, feet, bladder), and then awareness begins to focus so intently that the individuals are oblivious to even the passage of time. Flow theory is predicated on the notion that cues about time are gradually eliminated as one's focus on the task becomes more intense.

Sensory Deprivation

The second method we employ to make time stand still is *sensory depri-vation*. Flow states make time stand still by overwhelming our conscious bandwidth for processing sensory information so time cues fall away. Sensory deprivation creates timelessness in almost the opposite way: we actively strip away sensory cues (including time cues) so the brain cannot calculate how much time is unfolding. The best tool for this, which I have employed many times myself, is the sensory deprivation tank (SDT). It represents almost the opposite of a flow state because it depends on removing sensory information as the requisite step. In the SDT, the subject is immersed in a salty bath that is at body temperature. The salt content makes your body float weightlessly. The temperature is such that you cannot feel any difference or contrast between your skin and the outside world. There is no sound or light in the SDT.

When I first immerse myself in an SDT, my brain has to get used to having no incoming sensory signals. It panics a little. That's normal. I use meditative breathing to quiet myself down. Then it is like breaking through Earth's atmosphere. Finally, I am in deep space. I am floating. I am in the now.

I have nothing to preoccupy my brain except thought itself. Slowly, almost out of sheer desperation and frustration, my brain surrenders and leans back into the flotation of the water, and then it begins to hallucinate vividly. It is filled with music and shifting patterns of color and light streaming all over the place, and then, instead of floating, I'm flying. I can fly to any memory or scene or vision I can summon. And time seems to stand still. I cannot discern if I have been in the tank for ten minutes or ten days. The only indicator that would be reliable would be my bladder (which one scrupulously empties beforehand). So, that would be the only "internal clock" to which I might be able to refer.

Does Meditation Make Time Stand Still?

A brief note about meditation is warranted here. Many practitioners of mindful meditation have remarked on their sense of time being altered during meditative practice. A recent study evaluated test subjects who carried out mindful meditation and a control group who read poetry.

The study demonstrated that the meditative group was more likely to *underestimate* short intervals of time (lasting only a few seconds) while *overestimating* longer intervals (two to six minutes). The authors of the study concluded this dichotomy might be related to the study's methodology because subjects were queried about longer durations of time only after the meditation session was concluded. This led the researchers to surmise that the more focused the concentration of the individuals became, the more likely they were to feel that time was passing more quickly. This may suggest that the sensation that many meditation practitioners report of "an awareness in the now of the present moment" may resemble Csikszentmihalyi's flow states, where increasing focus is associated with a diminished awareness of the passage of time.

A Final Word About Time

The fluidity of time is not an arbitrary or esoteric notion of physics. Experiential time is an intensely personal property derived from our consciousness. In his novel *Kafka on the Shore,* the Japanese novelist Haruki Murakami has a character named Miss Saeki, a middle-aged librarian who warns the protagonist, Kafka Tamura, "Time's rules don't apply here. Time expands, then contracts, all in tune with the stirrings of the heart." The dynamic fluctuations of time we see in the brain are derived from our perceptions and emotions that lend our experiences of time their context, meaning, and propulsion. The most significant risk we face is not focusing enough on the rate at which time is flowing. Whenever we are genuinely engaged and absorbed by an undertaking, we shift into the realm of magic, where *time stands still,* held fast in instants of terrific concentration. And when our time is wasted, when we are involved in tedium, then our brains warn us to pay attention because *time is slowing down.* Finally, as we age, as the resource of time itself becomes increasingly precious, we see *time go faster.* To see how time flows is to take the pulse of our very lives as we live them.

On the Nature of the
Mind-Body Connection

All men dream: but not equally. Those who dream by night in the
dusty recesses of their minds wake up in the day to find it was vanity,
but the dreamers of the day are dangerous men, for they may act their
dreams with open eyes, to make it possible.
—T. E. LAWRENCE, *Seven Pillars of Wisdom*

Think of the latest blockbuster movies: Marvel's *Avengers, Harry Potter,* and *Star Wars.* These films grab us because they are stories of
ordinary people who somehow have gained extraordinary powers.
From the earliest myths, humans have wondered how to gain command
over the ordinary forces of the physical world with the extraordinary
powers of their mind.

The real story of the link between mind and matter emerges from
two separate areas of research: *psi,* on the one hand, and *psychoneuroimmunology,* on the other. Psi includes such unusual mental powers as
extrasensory perception (ESP), telepathy, and clairvoyance. The capabilities of psi can truly seem almost like superpowers. One researcher
called them "abilities well beyond the expected norms of human capacity, including control over mental and physical processes that transcend
the known laws of physics." Psychoneuroimmunology, on the other
hand, focuses on the intersection where the brain mediates the physical
and mental responses to stress and modulates the body's immune system. It is the ultimate connection between the "inside" of us—what we
feel and think—and the "outside" of us—where we may be invaded or

overwhelmed—and the body must mount sufficient defenses to provide us with the necessary protection.

Many of us have had direct experiences with both. We have encountered instances where we felt the strange effects of psi. And many of us have also witnessed the miracle of seeing the body being saved by the mind or the tragedy of its being destroyed by it. While such experiences may seem the exception, they're not. A Gallup poll found that 75 percent of Americans have had such extraordinary experiences.

This chapter is divided into two sections. The first explores psi and how our minds alter physical matter. The second examines psychoneuroimmunology—namely, what research reveals about the far-reaching effects of the mind upon the body's immune function. Separating the material into two sections is more a reflection of how the scientific work in these disciplines has taken place in two distinct academic silos over the last several decades. My concern is that this division risks creating the impression that there is a fixed, arbitrary border between psi and psychoneuroimmunology, when they actually represent a single veiled territory where the worlds of mind and matter freely intermingle.

Section I. Mind and Matter

Many years ago, I was lecturing on the scientific evidence about psi at my own University of Arizona. When I was done, I got on the elevator, and a fellow surgeon jumped in after me at the last minute. He proceeded to lambaste me for having "squandered a Grand Rounds lecture in the Medical College on such mental rubbish." Then he confided in me: "On the other hand, I have to tell you that I know immediately whenever anyone in my family is calling me on the telephone to tell me some really bad news—like a death in the family, or someone being diagnosed with cancer—I always know it beforehand because I have this overwhelming premonition of something dark and fearful enveloping me a minute or two before the phone is about to ring." He added, "I always know it and I'm never wrong about such things," and then stepped off the elevator at his floor. He turned, faced me, and shrugged, saying, "I have no idea about how or why it happens to me." I was dumbfounded, having just spent an hour lecturing on how such things happen. He left me at a loss for words.

The Brain Doesn't Work the Way We Think It Does

Before we launch into the nature of mind-matter-body connections, I confess that I and my neuroscientific colleagues have two glaring blind spots when it comes to insights into the functions of the central nervous system. The first is we think we have sound, logical ideas about the way *the brain should be organized* but we are almost always wrong. The second one is whenever we think we know *what the brain can do,* we underestimate its powers. And nowhere is this more apparent than when we explore the connections between mind and matter and mind and body.

Let me give you an example of the first kind of failure in our logic about how the brain should work. Our brains cannot escape the natural, anatomic logic of the "knee bone being connected to the thighbone." And that's how we anticipated the brain's cortex would be organized: our foot would be connected to our lower leg and that would be attached to our upper leg and so on. We expected that as far as the brain was concerned, it would receive sensory information from our bodies in one seamless map where one part flows into the next. But that isn't what at all really happened in the brain.

We know we were wrong because of the work of Wilder Penfield, a famous Canadian neurosurgeon. He assembled data from recordings he obtained in the operating room during brain surgery cases he carried out at the Montreal Neurological Institute. He would directly stimulate areas of a patient's sensory cortex while they were awake. He could then precisely map out what parts of the brain were devoted to sensation in what part of the body. I always want to remind readers that surgeons can operate directly on the brain in a patient who is fully wake because the brain is devoid of pain receptors. Penfield created a detailed schema from these intraoperative recordings of exactly how the brain had mapped sensation throughout the body. This became critical to guiding surgeons where they could operate in the sensory cortex and what part of the body might be affected if damage was not avoided.

Penfield demonstrated that the amount of sensory cortex the brain assigns to a particular part of the body is not proportional to the size of body part, as we once thought. Instead, he showed, our brains dedicate more sensory cortex to our thumbs or our lips than what is provided for the entire areas of our chest, abdomen, and pelvis combined. The regional anatomy is laid out by the brain in the cortex not by the size of the body part but by the importance the brain gives it. Not how we see it.

Contrary to expectations laid out in *The Skeleton Dance,* as far as our brain is concerned, the foot bone is *not* connected to the knee bone. As Penfield laid out his sensory map of the sensory cortex (called a homunculus; see figure 14.1), he discovered that the area devoted to our genitalia lies right under our big toe and sole of our foot. The trunk of the body does not end with the head. Instead, the brain has laid it out so it is capped with a hand topped by a gigantic opposable thumb, which looks as if it were about ready to become a float in the Macy's Thanksgiving Day Parade. Then comes the face and the head, cleanly decapitated from the rest of our body. And beyond the head, also lying in splendid isolation, comes our larynx and tongue. In short, the brain does not organize

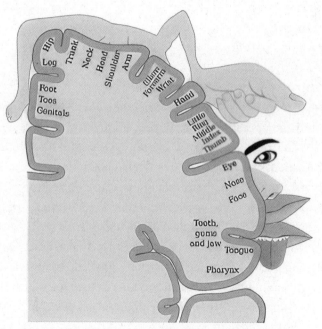

The homunculus is a 2-D representation of the human body as laid out in proportion to the amount of sensory cortex devoted to each area of the body.

the human body as one continuous structure but, instead, looks at it as a collection of warped, swollen parts, scattered like flotsam washed up on the beach. It's not the way we would have organized it, perhaps, but no one asked us, right?

The Quirky Nature of Psi

There is virtually no other place in neuroscience where we so consistently underestimated the brain's powers than happened in matters relating to psi. You'll want to pull your seat belt tight for what's coming. While you are buckling up, let me explain more about what psi is. Psi is the twenty-third letter in the Greek alphabet and stands for *p*. *P* as in *p* for "*psyche*," and "*parapsychology*," and the "*paranormal*." Psi takes in all manner of ESP, including clairvoyance and telepathy, because they all lie beyond our five physical senses. It also includes the area of *psychokinesis,* which is the ability to produce measurable changes in the physical properties of a material, a machine, or a computer system without touching it. Sit back and try to maintain an open mind because I am going to tell you about some real superpowers you might never have guessed you had.

Many of us have had strange and compelling experiences that we simply could not explain, or we know someone who has. Sometimes, they can surround a heartrending occurrence. Such was an incident first studied by British psychiatrist John Barker, a member of the Society for Psychical Research (SPR) , and then included in Sam Knight's book about Barker's investigative experiences [see REF A], titled *The Premonitions Bureau.* Members of the SPR had included Cambridge professor Henry Sidgwick, a renowned mathematician and economist. Other members were Edmund Gurney, a psychologist and one of the first people to investigate hypnotism. physicist William Barrett, Lord Rayleigh, and British prime minister Arthur Balfour [see REF B].

At the center of the premonitory event was Kathleen Middleton, who lived in the London suburb of Edmonton. Miss Middletone lived by herself in a small house, where she rented a room out to a lodger and offered lessons for children in both pianoforte and ballet, talents she had pursued vigorously since childhood. In the predawn hours of October 21, 1966, the fifty-two-year-old Middleton was suddenly awakened when a terrible sense of dread and foreboding gripped her. She felt herself struggling for breath, "choking and gasping and with the sense of the walls caving in." She stayed awake until Alexander Bacciarelli, the lodger, had returned from working the night shift, and she shared the terrifying account with him. He put a kettle on and boiled water so they could have tea.

While they were drinking their tea, a hundred and fifty miles away in the heart of coal mining country in Wales, lay the town of Aberfan. Outside the mine, a gigantic hillside of coal tailings had accumulated over years and had become supersaturated with weeks of rainfall. Unbeknownst even to the miners who worked there at the coal mines, the entire mountainside of coal tailings had been transformed into a gigantic, thick muck of coal and water. The black mountainside turned into a frightening, ponderous tsunami of over fifty thousand tons of coal slurry.

It raced down the mountain and the crashed into the valley below, burying dozens of houses and the Pantglas Junior School in a deadly, choking, inescapable mud of coal, dust, dirt, and debris over thirty feet deep. One hundred and forty-four people lost their lives that morning, including one hundred and sixteen children who suffocated inside the school. Dr. John Barker arrived the day after the disaster as hundreds struggled to dig the children's bodies out. The scene was reminsscent of the horrors of the Blitz in London during World War II. Seeing all the children's lifeless bodies being pulled from the wreakage he wrote: "The experience sickened me," He was so moved that he was keen to investigate any inkling of a preminiton of a tragedy on such a scale and how it might be avoided [see REF C] (138).

To understand how such experiences occur, we need to look at the first scientific inklings about psi that began nearly four centuries ago. They came from none other than the great mathematician and philosopher René Descartes (1596–1650). Early in life, Descartes had considered becoming a professional gambler. Instead, he went into engineering and analytic geometry. Descartes would become an empiricist philosopher and is considered the father of *dualism*. He articulated the overarching principle that *while the mind and the body appear to be distinct, one cannot exist without the other*. In essence, he was the first to frame the problem of the mind-body connection. Descartes, however, never gave up his love of gambling and recorded an unusual observation. It was his impression that happy, optimistic gamblers seemed to exert a more positive effect on seemingly random outcomes like the roll of the dice than did sad individuals.

Three and a half centuries would pass before researchers began to investigate the science behind how an individual could affect the outcome of random events. In the 1980s, the Princeton Engineering Anomalies Research (PEAR) laboratory was started by the dean of engineering, Dr. Robert Jahn. Instead of using dice or coin flips, the PEAR lab employed a computer, called a random number generator (RNG). An RNG generates the equivalent of a coin flip using the digital ones or zeros of its computer language to represent "heads" or "tails."

Psychokinesis

RNGs have two advantages: the first is they create a coin toss that cannot be physically influenced by human gimmicks, trick flips, or secret catches. No one needs to touch an RNG for it to work, and it is a reliable mathematical engine of true chance. Second, one can quickly generate thousands of trials with an RNG, and this gives scientists the statistical power to detect small but scientifically significant effects. The ability of a person's mind to affect random events is called micro-psychokinesis (micro-PK; from the Greek *micro* for "small," *psyche* for "mind," and *kinesis* for "movement"). At the PEAR lab, scientists accumulated indisputable evidence that some individuals had psychokinetic powers that could control the outcomes generated by an inert computer.

Volunteer test subjects were brought into the PEAR laboratory and placed either in a room with an RNG or in a room next door to one. The volunteers were asked at random intervals to exert no effect on the RNG, then to generate more zeros or more ones. Certain individuals were able to produce a substantial deviation in the output of the computerized RNG (see figure 14). These abilities were operator-specific, meaning some individuals had better abilities than others.

The PEAR lab evaluated the effects of micro-PK not just on other computer operations but also on mechanical devices like the motion of a swinging pendulum. One such experimental design involved an apparatus with more than nine thousand Ping-Pong balls that were released so they fell down and randomly collided with a series of pins and then ended up in one of nineteen different bins at the bottom of the device. The exact number of balls entering each bin was counted photoelectrically. Test subjects were asked to create a shift as to which bins the balls would fall into. To accomplish this, volunteers had to make the distribution of how the balls fell veer away from random chance. Three thousand experimental runs involving twenty-five different individuals demonstrated that the shift in the distribution of the balls was far beyond anything that was generated by random chance.

Recent research into micro-PK is using more sophisticated tools. The most recent iteration involved volunteers from three different countries who participated in the experiment at a distance. Their only connection to the experiment was via the internet to ensure test subjects were isolated from potentially confounding inputs from researchers or cues in the laboratory. Images were collected from a library of stock photos. One hundred photos of "positive prevailing mood" were selected. These were images of beautiful landscapes, cute animals, and so on. Another one hundred photos of negative images were selected; these depicted images of imminent danger or tragic catastrophe. The mood of the photos was reinforced by being displayed with consonant musical chords for the positive images and dissonant chords for negative ones, so the brain had multisensory modality input at each trial.

The images were selected by an RNG that used the splitting of photons from a laser beam as they struck a prism to randomly shuffle them. The images were then sent to the subjects on their computer screens or cell phones. Subjects were first tested without any intervention. Then subjects were taught to meditate to relax while they listened to positive affirmations like "I know everything is going to be just fine." The hypothesis was that as subjects felt greater optimism and positivity, they could exert a greater effect on making positive outcomes (that is, photos) manifest themselves. The experiments showed that there was, in fact, a significant effect of positive attitude on positive picture selection. This was solid scientific evidence of the impact of mood on the outcome of micro-PK. So, after the passage of nearly four centuries, Descartes's astute observations at the gaming table were proven correct. There is substantive scientific evidence that some individuals possess psychokinetic abilities.

The Effect of Psi on Biological Systems

It was only natural to wonder, could conscious intentional energy exert similar effects on biological systems? Carroll B. Nash, a psychologist at St. Joseph's College in Philadelphia, carried out experiments in which *Escherichia coli* bacteria were raised in tubes of culture medium and sixty college students were recruited and asked, at random, to either accelerate or retard the growth of bacteria. The result showed that normal volunteer test subjects with no prior psi experience or training could significantly increase or decrease the rate of bacterial growth at will. Other studies have similarly shown that test subjects are able to protect red blood cells against disruption and slow the growth rate of both leukemia and breast cancer cells in culture.

My own laboratory showed similar effects of intentional energy on the growth rate of brain cancer cells in culture. A genetic analysis of the affected cells showed dozens of individual genes in the cancer cells that regulated the rate at which cells proliferated had been profoundly affected by the exertion of conscious intentional energy. Our results suggested that such energies had a direct effect on the cancer cell's nucleus and influenced the regulation of gene activity in the cell's DNA. I know. I was so skeptical I insisted my lab manager reproduce the experiments no fewer than four separate times. And every time, the only tumor cell cultures that stopped growing were the ones where world-class energy healers had set their minds' intention against the tumor cells being able to replicate. It's a bit mind-blowing that the human mind could exert volitional effects over a wide array of living cells. I want to point out that this series of experiments employed the talents of very experienced energy healers (some of the best in the country) and they were able to affect cell growth on culture plates under highly controlled experimental conditions. I want to caution readers that I and the other researchers do not mean to hold out that such techniques are a mainstream treatment for an individual stricken with cancer. But it's true data and helps us to grasp a deeper understanding of the power of the mind-body connection.

Remote Viewing at a Distance

I have saved the most startling demonstrations of the mind's "superpowers" for last. Besides their groundbreaking work on micro-PK, PEAR lab investigators spent more than a decade and hundreds of experimental trials evaluating so-called remote viewing at a distance (RVD). RVD is a demonstration of a kind of telepathy or ESP. In the case of the PEAR lab experiments, one subject was selected as *the viewer* (or *sender*) who would be assigned to look at a particular target site at an undisclosed geographical location. Another subject was selected as *the receiver* (or *percipient*) who was assigned to describe whatever impressions of the target sites they might receive from the viewer. There was no communication between receiver and viewer. The target site selected for the viewer

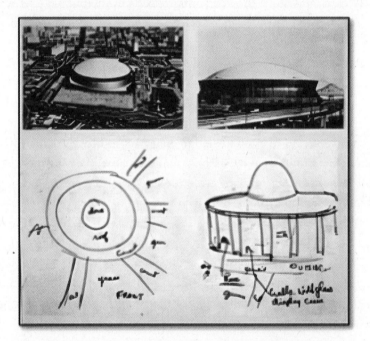

Remote viewing at a distance: The two handmade drawings were generated by two different percipients to document the impressions they had obtained telepathically from viewesr at the actual target site. The photograph is what the actual target looked like to the viewer. The features that were identified by the percipients are striking. There was never direct communication allowed between the percipient and the viewer during the course of the experiments so all impressions were telepathic.

was never disclosed beforehand and was randomly assigned by a computer to the individual viewer. The viewer was then called upon to spend fifteen minutes observing features of the target site. It could be visited in person, or it could simply be viewed as a photograph of a particular target site. In the meantime, the percipient would be asked to record his or her impressions of the sender's viewing of the target site. Below are examples of photographs of the actual remote site and corresponding drawings made by the percipient:

As you can see, receivers could exhibit an uncanny ability to generate an accurate impression from the sender. As hard as this may be to understand, the results have been replicated in dozens of laboratories by different researchers. In 1988, an analysis of more than twenty-six thousand RVD trials concluded that the likelihood of such results being due to random chance were less than one in 10^{20}, or 1 followed by twenty zeros. In 1996, Congress commissioned an independent assessment of the evidence for RVD. The final report stated,

> Using the standards applied to any other area of science, it is concluded that psychic functioning has been well established. The statistical results of the studies examined are far beyond what is expected by chance. Arguments that these results could be due to methodological flaws in the experiments are soundly refuted. Effects of similar magnitude have been replicated at a number of laboratories across the world.

There is one detail that I have saved for last because it is, in a way, the most mind-blowing of all the results to emerge from the RVD experiments. In many cases, viewers were asked to make a drawing of the target site but they had no idea they were draw a target a day *before* the photograph would be randomly assigned to a sender! In other words, the receivers were being asked to draw a target site long before the sender would ever know what their designated target would be or where it was located! And many times the viewers were still able to accurately draw impressions about the site before it was ever known to anyone. This phenomenon is known as precognition (that is, "knowing before knowing"), and it is a prominent feature of psi that defies imagination but has been demonstrated repeatedly since the PEAR lab experiments.

We have seen that psi defies logic. But it also flouts the normal rules of Newtonian physics. We have seen that future site designations could actually shape the RVD readings in the present. Investigators also found psi does not function like traditional energy sources. For example, the distance between sender and receiver does not affect the results. Also, in contrast to typical energy sources (like radio waves, X-rays, and infrared), psi is not blocked by isolating the sender or receiver deep underground in a mine or cave. Nor is the transmission blocked by electromagnetic or lead shielding. Whatever powers psi conveys appear to disregard the rules that govern other electromagnetic forces operating in the universe, like gravity, electricity, thermal energy, or light. Nonetheless, the existence of psi has been proven repeatedly over the course of more than half a century. And, we have learned from research into seemingly anomalous results, psi can open up whole new areas of inquiry like quantum mechanics.

I admit that at first I was a skeptic about psi. But as a scientist I began to seriously reconsider the field after hearing a detailed review of the PEAR data when I was an intern. The more I delved into the peer-reviewed articles on the anomalous findings, the veracity of the hard science seemed to hold up. In the introduction, I mentioned that Einstein could not bring himself to believe in some of the "spooky" conclusions that quantum mechanics predicted, like entanglement, where typical rules of energy transmission and light speed do not apply. Nonetheless, those conclusions were proven to be true. And so far, psi has also proven to be true. Still, the mechanisms of psi need elucidation, but there is abundant circumstantial evidence to support both precognition and remote viewing.

The Neuroscience of Psi

We cannot help but wonder, what in the brain makes psi possible? Scientists are using the latest technologies to find out. Electroencephalography evaluates the frequency, amplitude, and wave patterns of electrical activity in the brain. The earliest work evaluated the linkage between the sender-percipient pair. The results showed that when a viewer was in "sending" mode, there was activation in his or her occipital lobe. This is where one would naturally expect an image to be processed by the *viewer*. But there was also coincident EEG activity occurring at the same time in the receiver's visual cortex as the thoughts of the images were being transmitted to him or her. These studies showed the brains of the sender and the receiver were linked together with the same parts of the brains being activated in both. They have been replicated more than a dozen times. A later iteration used EEG and fMRI imaging in combination. Same results: reciprocal activity in both brains on EEG analysis and with fMRI-based imaging. So, activity in one brain simultaneously evoked shared activation in the other. Cerebral entanglements.

Two famous mentalists, Gerard Senehi and Sean Harribance, are renowned for their psychic and telepathic abilities. In Mr. Senehi's case, he was asked to reproduce line drawings that were being displayed at random intervals to a control subject who was functioning as the sender. During the telepathic task, Mr. Senehi showed marked activity in his *right* hemisphere, especially the *posterior visual area* and the *parahippocampus,* which is a bridge between the memory areas and the sensory perceptions. There was also activation in a remarkable, small area adjacent to the visual cortex, called the cuneus, which I will get to in a moment.

In 2012, Mr. Harribance was also evaluated using very accurate EEG analysis by a well-respected research group in Ontario, Canada. Like Mr. Senehi in the earlier study, Mr. Harribance showed greater activity in the *right parahippocampus* during his "intuitive" or telepathic states.

An unrelated study used neuroimaging to evaluate brain activity in distant intentionality (DI; that is, sending thought at a distance) in eleven healers who participated regularly in healing patients at a distance. The healer was asked to direct his or her DI for random two-minute intervals. Again, researchers found that there was increased activity in our old friends the *anterior cingulate area* (ACA) and the *mesio-orbital frontal lobe,* as well as that not so familiar part of the brain called the cuneus.

As you will recall from earlier chapters, the MOFL provides executive thinking and risk-benefit and reward-punishment analysis. The MOFL sends these thoughts to the ACA, where emotion arises because risks and benefits also have emotional and behavioral consequences. It also adds a very significant motivational component to our behavior. And, finally, there is the cuneus.

We saw the cuneus implicated in chapter 1 in the discussion of consciousness. The cuneus is a small wedge-shaped (*cuneus* means "wedge" in Latin) area of cortex, lying between the visual and the sensory areas on the brain's surface. It is involved in sensory processing, but it also has a unique function: it recognizes our selves as distinct from everyone else. That may seem like something we would take for granted but, again, the brain does not work the way we think it should. The brain creates an internalized experience of a unified self. Damage to the cuneus has been implicated as part of a clinical state called alien limb syndrome in which a patient cannot recognize his own limb as belonging to himself and instead sees it as someone else's extremity.

The cuneus is also thought to be an important contributor to our sense of identity according to the theory of the mind (TOM). Thomas Nagel tried to address the notion of self-identity and awareness in his famous essay titled "What Is It Like to Be a Bat?" In it, he makes the case that while we might want to imagine ourselves as a bat, our only frame of reference is constrained by our subjective experience of being human. Nagel points out that we cannot imagine a bat's sense of the world conveyed to it by its sonar because it is a form of perception we do not possess. TOM may also play a role because it contributes to our notion of intentionality, of the desire to express our personal will to see an objective realized.

There is an intriguing anecdotal report—remembering that *all* medical science begins with the publication of anecdotal cases—on two patients with extensive frontal injuries. MRI brain imaging studies in both these two patients showed they had suffered significant damage and loss of brain tissue specifically in the mesio-orbital frontal lobe. However, when these two subjects were asked to change the outcome of an RNG, they were significantly *more* capable of exerting effects than a normal control subject. Results like these suggest that looking at exceptional individuals (like healers or mentalists) can obviously show us what a small minority of gifted individuals can do. But maybe we should ask, what is holding the rest of us back? Rather than ask what is going on in the brains of these exceptional individuals, we may need to analyze what is inhibiting psi from occurring in normal subjects, like you and me? This raises the possibility that *the frontal lobe may normally inhibit us from exercising our psi capabilities.* It may suggest that we need to *think less* and *feel more* to investigate the peculiar and anomalous features of psi. This may be one of the reasons that mindful meditative states may make one more susceptible to developing one's capacity for psi.

I have spent many hours pondering the role of intuition in my own life over the years. I have personally experienced several situations where my own irrational and incomprehensible intuition—being a big fan of comic books, I like to call it my "Spidey sense"—has saved my life. One of the most dramatic and meaningful (because it was also witnessed) occurred in the Swiss Alps. I was hiking down with my wife after spending the night at the Great St. Bernard Hospice, a beautiful little inn that straddles the third-highest mountain pass in all of Switzerland at eighty-one hundred feet above sea level. We had decided to hike down to the town of St.-Rhémy, almost four thousand vertical feet below the St. Bernard Pass. There were two trails down. One would be the fast way and save us a lot of time, and we were eager to catch a train. And the other was a much longer way down and more time-consuming. We got to the junction point where the two trails split off, and my wife, Jane, naturally headed toward the shorter, faster route. Suddenly I had a feeling as if someone had walked over my grave and I said, "No, not that way. The other way."

My wife looked at me quite puzzled and asked, "How come?"

I looked at her and said, "I don't know why. I just have a bad feeling."

So, we turned and took the long way down. About an hour and a half later there was suddenly a big landslide where a gigantic mass of melting snow broke loose and took hundreds of tons of loose rock with it. It was right where we would have been had we taken the other trail.

Jane just looked at me a bit spookily and said, "Well, now we know why you had a bad feeling."

To this day, I can't explain it. But it has happened so many times in my own life I've learned not to ignore it or question it. I simply try to just let it register and go with it.

I have made a relatively in-depth evaluation of psi for the reader because I believe there are several important points that emerge from this analysis to keep in mind:

1. The brain has profound powers beyond what we ever guessed. Psi is a good example: it is a power with which the mind can affect machines, matter, and living cells and tissues.
2. These powers defy strict Newtonian physics but have several properties that are reminiscent of quantum mechanics.
3. Precognition plays a role in psi, although how information about future events can be consistently relayed to the present is unclear.
4. Psi transmission between a sender and a percipient appears to require a functional, reciprocal connection between their two brains.
5. Areas of the brain that have been implicated by EEG and neuroimaging studies in the generation (and reception) of psi appear to be preferentially right-hemispheric functions.
6. Psi includes the sensory cortex (especially the visual areas), ACA, MOFL, the parahippocampus, and the cuneus.
7. The role of the cuneus in psi is noteworthy because the cuneus is thought to play a role in the TOM believed to generate conscious awareness.
8. Our normal cognitive pathways may actually interfere with our ability to feel and generate psi, and mindfulness training may assist in overcoming such inhibitory influence.

Section II. Psychoneuroimmunology

The Mind and the Body

The properties of psi set the stage for grasping the power of the mind-body connection in disease states and in establishing wellness. But before I delve into *how* the brain exerts its effects, I want to turn to *what* the brain can do. To do that, we need to answer two questions. First, can your brain kill you? And, second, is there a way your brain can save you?

Walter B. Cannon

One of the discoveries about the effect of the mind on health outcomes came from the battlefields of World War I. Walter Cannon, the precocious and brilliant Harvard physiologist, had tended to American soldiers at the front. He discovered that wounded soldiers went into shock from losing too much blood and would then go on to suffer cardiovascular collapse and die. He noticed that before this collapse occurred, the soldiers' blood became acidic from a buildup of excess lactic acid in the bloodstream.

Cannon had seen a similar picture years earlier in cats in his own laboratory. He noticed that cats that had endured several weeks of stressful living conditions that resulted from being housed in the same facility with a lot of barking dogs also exhibited elevated acidic blood. He eventually demonstrated that this acidosis could be traced back to stress-induced hormones pouring out of the adrenal glands—epinephrine, norepinephrine, and corticosteroids—as part of the cat's fight-or-flight mechanism. Cannon had also observed that the cats that were housed the longest with the dogs would lose weight, develop non-healing skin ulcers, and go on to die from systemic cancers. But why, he wondered, was stress so devastatingly injurious and even fatal?

Voodoo Death

Cannon would find his answer in the jungles of the Caribbean and Africa. He had come across one of the darkest and most dramatic manifestations of acute stress the mind-body connection could create. He published his observations in a startling research paper that is still considered one of the landmark, classic papers ever published in physiology. He titled it " 'Voodoo' Death," and it appeared in the prestigious journal *American Anthropologist* in 1942. In it, he cautioned his reader,

> Before denying that "voodoo" is within the realm of possibility, let us consider the general features of the specimen reports. . . . First . . . is the fixed assurance that because of certain conditions, such as being subject to bone pointing or other magic, or failing to observe sacred tribal regulations, death is sure to supervene. This is a belief so firmly held by all members of the tribe that the individual not only has that conviction himself but is obsessed by the knowledge that all his fellows likewise hold it. . . . The question which now arises is whether an ominous and persistent state of fear can end the life of a man. . . . The suggestion which I offer, therefore, is that "voodoo death" may be real,

and that it may be explained as due to shocking emotional stress—to obvious or repressed terror.

What Cannon proposed in "'Voodoo' Death" was a radical idea; namely, what thoughts occurred in your head had the power to kill you. If it had been anyone other than Walter B. Cannon who had proposed it, the idea would have been dismissed. He hypothesized that the voodoo curse caused so much anxiety and fear in the individual that it caused a massive outpouring of adrenocortical steroids until the depleted adrenal glands collapsed and the individual went into shock and died. It was an accelerated version of what he had seen in the laboratory cats.

I do not think anything could have prepared me for the alarm I felt at seeing "voodoo" at work with my own eyes. In my earlier book *The Scalpel and the Soul,* I told a brief story about a nurse who accompanied me to a Fang fishing village located deep along riverways of the Ogooué River in the Gabonese jungle. A hut had been specifically constructed by the elder women of the tribe for the ritual teaching of young women who had experienced menarche for the first time. So powerful was the collective magic of these women elders—and of the hut that they had constructed—that any man who dared to even touch the hut would die. This prohibition was necessary to avoid having men "peek" into the hut to scout out potential brides.

In our party, we had one native Fang nurse and he wanted to prove to us—the white Europeans and Americans—that he was a man of the world, sophisticated, and above all this primitive, irrational nonsense. He decided to make a big show of bravado by touching the sacred hut with his foot. He stepped on the corner of it many times and then said, "Tu vois bien. Rien ne m'a attacqué ni blesse!"* Well, he might have wanted us to believe that he was above all that superstition, but he did not believe it himself. The next day he became deathly ill. And despite my giving him intravenous fluids, steroids, and antibiotics, and tending to him myself night and day, he died three days later. Of what? Magic? Superstition? Really, just his own beliefs. Whatever it was, he was very dead from it. Voodoo death has presented tantalizing clues to the mind-body

---

* "You see, nothing attacked me or hurt me."

connection.

Voodoo on Wall Street

But you don't need to be in the African jungle to believe in *juju and voo-doo*. It happens in concrete jungles too. Between 1985 and 2000, Kenneth Lay had made the Enron Corporation into the second-largest energy company in the world. By the turn of the twenty-first century, Enron was worth more than $60 billion and ranked as one of *Fortune* magazine's "Most Admired Companies." In the late 1990s, Lay became a close friend of many politicians, including the Bush family. When George H. W. Bush flew to Washington, D.C., to be sworn in as the forty-first president of the United States, it was in Ken Lay's private corporate jet. But Enron's mojo was too good to be true, and rumors began to circulate about unethical and illegal accounting practices. A series of investigations were eventually launched by the Justice Department, and Enron's stock fell from $100 to a few pennies per share. Lay, now bankrupt, was found guilty in federal court on six counts of securities and wire fraud, punishable by up to forty-five years in prison. By all accounts, Lay was beside himself with anxiety and worry. He died of a heart attack just days before he was to go to prison. Sometimes, the heart passes its sentence before a judge can. But where the heart can condemn, it can also save. One of the compelling pieces of evidence of the positive impact the mind can have on outcomes is the placebo effect.

Anzio and the Birth of the Placebo Effect

The Battle of Anzio in Italy was one of the bloodiest in all of World War II. During some of the heaviest fighting, the Allies had run critically low on supplies, in part because German bombardments had specifically targeted the Allied medical supply depots. An anesthesiologist, Henry Beecher, was preparing incoming wounded for triage, lining up those who most urgently needed surgery. Beecher had been the anesthetist in chief at the Massachusetts General Hospital before World War II. His job was to ensure that there was adequate intravenous access for fluid resuscitation to guard against shock, and to give the soldiers a big slug of morphine before they headed directly to the operating room.

As one soldier after another streamed into the field hospital on stretchers, Beecher was suddenly informed that there was no more morphine left anywhere along the invasion beachhead. He was thunderstruck. Without morphine, there would be no way to get soldiers readied for the surgery they so desperately needed. But then, as he looked on, a nurse very deliberately drew up saline into a glass syringe. The nurse leaned over and told the soldier that she was going to give him a big shot of morphine and "that within seconds he was going to feel very, very drowsy." To Beecher's utter amazement, the soldier became sleepy, his eyes fluttered and then closed; then his heart rate slowed down as he became sedated and stabilized for the operating theater.

When Beecher returned to civilian life, he did not forget the lesson the nurse had taught him. In 1955, he published an influential paper titled "The Powerful Placebo." The word "placebo" is derived from the Latin verb *placere,* meaning "to please." Beecher led substantive efforts to ensure that pharmaceutical research was conducted in placebo-controlled trials to demonstrate their effectiveness in a fair and impartial manner. Beecher wrote prophetically, "Preservation of sound judgment both in the laboratory and in the clinic requires the use of the 'double blind' technique, where neither the subject nor the observer is aware of what agent was used or indeed when it was used. This latter requirement is made possible by the insertion of a placebo, also as an unknown, into the plan of study."

Placebo: Believing Is Seeing

Nowadays, the pharmaceutical industry spends an estimated $2 billion a year trying to account for the placebo effect and rule it out in the evaluation of new clinical trials. In 1978, researchers at the University of California, San Francisco, investigated the role of placebo-induced analgesia in routine extraction of wisdom teeth. They divided patients into three groups: (1) received morphine without being told by the physician; (2) received morphine and were informed that that drug was being given; and (3) received an inert saline injection but were informed that it was morphine. Groups 2 and 3 had the same level of pain relief as the patients in Group 1 who had received up to twelve milligrams of morphine. That is a pretty large dose of morphine.

The effect of the placebo could also be reversed if they gave the powerful opiate antagonist naloxone (trade name Narcan), meaning it was the brain's own endorphins that were responsible for the placebo's analgesic effect. PET scans on the placebo studies show that an analgesic placebo works precisely because it activates the same centers in the brain that are stimulated by exogenously administered opiates. In short, if we believe we are receiving a narcotic—even if it's just saline—then our brain acts on it.

The Placebo Conundrum

Dr. Francis Moore, the surgeon in chief at the Peter Bent Brigham Hospital at Harvard Medical School, once confided in me: "There is no more important confounding force in all of medicine than the patient's own earnest desire to attend to a positive outcome." What he meant by this is that *patients want to get better.* They want to believe that the intervention—be it a pill or a surgical procedure—is going to help them. And it can skew a physician's assessment when they merely ask their postoperative patient, "Are you feeling better?" That's because the patient is so motivated to to feel better. So, of course, they're far more likely to say, "Yes, I do feel better." This means the physician, in particular, has to pay close attention: "Did my intervention really make my patient measurably—objectively—better?" And doctors—surgeons, especially—have a notorious bias in favor of their own work. There is an old adage that if you say to a surgeon, "Name the three best surgeons in your field," they always have difficulty coming up with the other two names.

In Finland, however, a group of orthopedic surgeons decided to undertake a daring experiment to answer the question, "How good is the outcome of my surgery?" Routinely, these surgeons would see a lot of patients every year who were suffering with a tear in the meniscus (the cartilaginous pad between the upper and the lower leg inside the knee). Such tears are diagnosed with MRI, and the common treatment is to make a small incision for the placement of an arthroscope and remove the damaged edge of the meniscus, a procedure called an arthroscopic partial meniscectomy (APM).

For this study, patients were randomized into two groups: one group underwent a real APM, and the other group had a sham surgery performed so, when they woke up, they had all the incisions from having undergone an APM, but only the first group had real surgery performed. As an aside, I can tell you that such an experiment with sham surgery would have had a tough time getting clearance today in the United States. With two years of follow-up in both groups, the authors concluded that there was no detectable differences between patients who had actually undergone APM and those who had had sham surgery performed. This is true not only in orthopedics. So, placebo-controlled clinical trials are a must. But the introduction of placebo controls is not without its own kinds of problems.

A dramatic example of the power of placebo was reported by Dr. Maj-Britt Niemi in her article "Cure in the Mind":

> A man whom his doctors referred to as "Mr. Wright" was dying from cancer of the lymph nodes. Orange-size tumors had invaded his neck, groin, chest and abdomen, and his doctors had exhausted all available treatments. Nevertheless, Mr. Wright was confident that a new anticancer drug called Krebiozen would cure him. . . . Mr. Wright was bedridden and fighting for each breath when he received his first injection. But three days later he was cheerfully ambling around the unit, joking with the nurses. Mr. Wright's tumors had shrunk by half, and after 10 more days of treatment he was discharged from the hospital. . . . He remained healthy until two months later, when, after reading reports that exposed Krebiozen as worthless, he died within days.

The physician's dilemma with the power of placebo was poignantly illustrated by a case of Dr. Isadore Rosenfeld's. Dr. Rosenfeld was a beloved and erudite cardiologist, and a powerful voice in support of integrative and alternative medicine. He reported on a patient of his who was suffering symptoms of moderate angina. The patient only ambulated in his apartment slowly with a cane, could not go for walks in the park, and did not even have the stamina to play chess in the park with his grandson. Dr. Rosenfeld was asked to evaluate a new medication that would be tested in a randomized double-blinded study. He started the patient on the new pill and very quickly saw the man's exercise tolerance (on the treadmill) improve dramatically. He put his cane away and began walking in the park. His EKG showed improvement. He had no more anginal symptoms. Rosenfeld was pleased and he felt confident, from the clinical responses he was seeing, that his patient had received the active, new medication.

At the end of many weeks, the laboratory investigation came to an end, and Rosenfeld received an envelope from the manufacturer that "broke the code," as it is called in clinical medicine. The letter informed him that his patient had received nothing but placebo, an inert sugar pill. So, what should Dr. Rosenfeld do? Remember Dr. Moore's admonition: "There is no more important confounding force . . . than the patient's own earnest desire to attend to a positive outcome." So, should Dr. Rosenfeld have said, "The pill didn't do anything for your heart. It was a sugar pill. Totally inert. It was all in your head." Or, should he have handed him another bottle of placebo pills and said, "Here's some more medication for you. I'm pleased with how well you are doing. Continue taking these and you will continue to feel better. Keep going for walks in the park." I won't tell you what Dr. Rosenfeld did, but I can tell you what I would have done: I would have kept writing him prescriptions for sugar pills. *In the end, the placebo effect is proof of the substantive power of the mind's ability to regulate the body.*

Psychoneuroimmunology: A New Brain Function

At the time, Lay's heart attack death days before his sentencing grabbed major headlines. Now it is widely accepted that sudden cardiac death can often be related to chronic overstimulation of the sympathetic nervous system—precisely how Cannon had hypothesized in voodoo deaths. More recently, brain imaging has shown how the mind can exert powerful control over the mechanisms that underlie disease.

The Brain and the Stress Response

There are *two powerful pathways* by which the central nervous system can directly impact our health. The first is through the modulation of physiological (and emotional) *stress*. And the second is through control and amplification of *inflammation*. In fact, it is a little artificial and arbitrary to talk about them as independent systems because they are congruent ones as you'll see.

We already discussed Cannon's landmark findings on stress and activation of epinephrine, norepinephrine, and cortisol. Since then, we know that the brain is intimately involved in stress and the fight-or-flight response. The brain is responsible for revving the adrenocortical response through activation of the sympathetic nerve fibers. It also determines the output of adrenocorticotropic hormone (ACTH) from the hypothalamus of the brain. ACTH is the master peptide that orchestrates the axis between the hypothalamus, the pituitary gland, and the adrenal glands, the so-called hypothalamic-pituitary-adrenal axis. The hypothalamus is the center of the brain that regulates the sympathetic nervous system and, thus, regulates the output of adrenaline and noradrenaline. Immediately below it lies the pituitary gland, the master hormonal control center of the whole body. It also secretes ACTH, which targets the adrenal glands to secrete cortisol.

The HPA axis is the link that was missing for Cannon about how an event in the world around us (like the barking of a dog) could get transmitted via our senses to the brain and how that, in turn, could produce a massive outflow of hormones in the body as part of the fight-or-flight response. *This is the primary mechanism of how the body mounts a response to stress.* The problem in modern-day society is not that we mount a response to stress. The brain is well designed to produce a full-scale hormonal activation to an acute threat. Our modern-day problem is that we no longer face acute stress. Instead, we live in an environment of chronic (and, in many cases, near-constant) stress: a condition we can adapt to only poorly.

The Brain and Immune Function

The problem of our response to chronic stress is twofold. The first is the feedback loop in the brain that controls the HPA axis. When it sees chronically elevated stress-related hormones, it tries to turn down "the gain" on the system by blunting the responsiveness of the adrenal gland as the stores of hormones in the gland become depleted. Chronic stress therefore leads to a poor stress response.

The second is that the HPA axis is intimately linked to the response of our immune system. In the past, doctors were taught that the immune system and the CNS were two independent systems. We now know that the linkage between the mind and the body is a substantive, two-lane highway that allows the brain to closely monitor the body's inflammatory condition and directly modulates the state of its inflammatory response.

Recently, brain imaging has revealed that the amygdala, hippocampus, hypothalamus, striatum, and midbrain all mediate our inflammatory responses. These areas also govern and modulate specific emotional responses. It seems likely that not only is the individual's state of mind affected by conditions that activate widespread inflammatory responses and illnesses but the converse is also true. The individual's emotional state can modulate his or her neuroendocrine and autonomic output, thus affecting the brain's direct control over inflammatory reactivity.

Inflammatory cytokines are a good example of this. They are a series of chemical messengers secreted into the bloodstream to govern the body's immune response. So, for example, if you were to have an infection in your hand, then cytokines would quickly flood into that area. Some of the best known and studied cytokines are interleukin-1 (IL-1), interleukin-6 (IL-6), and tumor necrosis factor-alpha (TNF-α). Typically, we think of cytokines as being released by the immune-mediating cells in the bloodstream when they confront an antigenic or allergic challenge. But, as has happened countless times before, scientists underestimated the "interest" that the CNS takes in monitoring such responses.

First, cytokines make an area, like your infected hand, hurt. That helps the individual to "splint" the affected part of the body and reduce movement to help slow the spread of bacteria. As the infection worsens, more cytokines are summoned. This produces more swelling in the hand and, eventually, fever. By producing swelling, the cytokines facilitate the exchange of oxygen and chemicals and rev up metabolism in the area, so the chemical reactions are accelerated. By mounting a fever, the brain will make the affected individual slow down his or her overall activity and thus reduce unnecessary energy expenditures. At the same time, raising the body's temperature helps accelerate chemical reactions in the affected limb, like your hand. As you know, if the infection goes on long enough, the hand also becomes warm and red. This occurs because the cytokines have dramatically increased blood flow to the infected limb, again to help accelerate delivery of white blood cells to combat infection.

The CNS has what is called "a privileged status," meaning it is protected from chemical and cellular events happening in the body by a kind of molecular "firewall," called the blood-brain barrier (BBB), built up around the brain's blood vessels. But there are a handful of places in the brain where the BBB is conspicuously absent. The brain has these sites so it can specifically monitor external metabolic and chemical changes going on in the body's environment. One of these areas is the hypo-thalamus, where we see this kind of "window" on the body's status. Here, the cytokines circulating in the body can leak in and directly interact with neurons in this area. The brain cells, in turn, can activate important metabolic pathways, like the release of glucocorticoid steroids, when the individual experiences stress.

During an illness, the stricken individual will often feel fatigued and listless and lie still in bed almost in a depressed condition for days to limit energy expenditure. In experimental animals, these same behav-iors can be induced by directly exposing the hypothalamus to cytokines. Circulating IL-1, IL-6, and TNF-α all appear to directly diminish activ-ity in the anterior cingulate area, the basal ganglia, and the dopaminer-gic center, the nucleus accumbens. This is important because reduced activity in the ACA is associated with depressed affect. The inhibition of the dopaminergic pathways (nucleus accumbens and basal ganglia) may help reduce appetite and motivation to undertake tasks that might inadvertently expend valuable calories needed to combat illness. In fact, a patient's clinical improvement can be predicted by the diminished effect of inflammatory markers on these specific areas of the brain.

Meditative Mindful Practice and Immune Function

There has now been ample study of the application of mindfulness prac-tices in disease states. In a cohort of 250 women suffering with breast cancer, the subjects received more long-term improvement in mood, anxiety, and overall quality of life from meditative mindfulness practice (MMP) than a matched cohort of women who underwent supportive group psychotherapy. These positive results of MMP lasted for more than a year after the intervention. We know that the MMP is accompanied by improvements in self-reports in a host of disease states. The questions arises: Is this all psychological, or is there a biological component too?

There are now several validated studies that MMP directly impacts immune function. C-reactive protein (CRP) is a blood-borne marker of the body's general, systemic condition. It can be elevated in a host of inflammatory conditions ranging from bacterial infection, hepatitis, heart disease, inflammatory bowel disease, rheumatoid arthritis, and cancer, among others. Three studies have shown MMP reduced CRP levels in patients suffering from breast cancer, ulcerative colitis, and rheumatoid arthritis. In addition, MMP improved cell-mediated immunity (as measured by CD4 and T-cell counts) in patients undergoing treatment for human immunodeficiency virus and in a separate group of patients being treated for breast cancer.

MMP has been shown to improve immune biomarkers in individuals not suffering disease. For example, isolation and social loneliness are associated with depression and a fall in immune function. MMP improved immune function in this group. It even improved antibody responsiveness to annual flu vaccination among healthy corporate employees. It is postulated that MMP is associated with an overall improvement in well-being and optimism and that enhanced immune function may be secondary to elevation in mood and attitude. One recent study, however, has suggested that improved processing of inflammatory markers by brain cells lying in those specialized areas where the BBB is missing may help augment immune function as a direct effect of MMP.

I am struck by how easily and readily the field of medicine embraced the idea that stress kills. As doctors, we find no problem embracing "the Dark Side," so to speak. Sure, stress can make your heart stop. It can make you age prematurely. It can make you fat. Demented. Stroked out. Tremulous. Impotent. Blind. Dumb. But we must also acknowledge that while negative emotions can lead to bad medical outcomes, positive emotions can lead to good ones. There is substantially less research written about positive results. The irony is that you can name almost any big, bad medical syndrome and I can practically guarantee you I can come up with four well-designed, placebo-controlled, peer-reviewed articles showing you that negative emotion can cause it. And yet there are still so few studies looking at the positive side of the coin. The placebo is one inescapable demonstration of the positive power of the mind over the body.

The findings about the mind-body connection are important. Let me sum them up as follows:

1. It is difficult to understate the influence of the mind-body connection in wellness and illness.
2. The negative aspects of the mind-body connection are powerful enough to cause an individual to die as a direct and immediate result of acute stress.
3. Chronic stress leads to the depletion of adrenal reserves and dysfunction of the normal feedback system in the HPA axis that controls our response to stress.
4. The placebo effect is one of the most powerful, dramatic, and confounding manifestations of the mind-body effect.
5. The administration of a placebo for analgesia shows that the brain re-creates a pattern of pain suppression that is nearly identical with that evoked by administration of opiates.
6. Placebo effects have called into question the efficacy of dozens of pharmacological and surgical interventions.
7. The brain exerts direct control over the body's immune response and can modulate the body's inflammatory biomarkers.
8. Effects of MMP go beyond just elevations in mood or sense of well-being and extend to specific, beneficial modulation of immune responses in a host of clinical conditions.

For too long, the field of medicine has sought to ignore or belittle the power of the mind over the body. There is no other field that so celebrates ideological conservatism and enshrines its own dogma with the ferocity seen in medicine. And it will defend those positions with such zealousness that those who dare question these positions will often find themselves the victims of derision, harassment, and even outright expulsion. Ignaz Semmelweis was the great champion of nineteenth-century obstetrics who advocated for rigorous hand washing and disinfection as a way to reduce the spread of infection from one delivery to the next. He was chased out in disgrace from no fewer than three countries after demonstrating the clinical efficacy of his antiseptic methods in three separate studies. He was driven crazy by the fact that no would heed his advice and save mothers from unnecessary infections, and he died in an insane asylum. The surgeon who first attempted to demonstrate the clinical feasibility of laparoscopic surgery in humans was immediately expelled from his regional surgical society as soon as he completed his first procedure. The two scientists who proved that peptic ulcer disease was caused by a bacterium, *Helicobacter pylori,* were not even allowed to present their findings at any scientific meeting and could not get their manuscript accepted for publication in a peer-reviewed medical journal for more than two years.

Two decades later these same two doctors would be awarded the Nobel Prize in Medicine for their pioneering discovery. Medicine is the sanctuary of unrepentant skepticism. The profound issues of the mind-body connection are only now beginning to be addressed in the mainstream thinking of medical education and training. In the meantime, every patient and family needs to take these studies and their conclusions to heart and remember that it has taken science more than four centuries to acknowledge what many of us have always sensed intuitively: what we feel in our inner thoughts has the power to shape events in the outside world. We are also discovering the outside world can directly affect our state of mind and immune function.

Nature Deficit Disorder

Nature deficit disorder (NDD) was coined by Richard Louv, author of *Last Child in the Woods,* to describe a condition among children characterized by "diminished use of the senses, attention difficulties, conditions of obesity, and higher rates of emotional and physical illnesses. Research also suggests that the nature-deficit weakens ecological literacy and stewardship of the natural world. These problems are linked more broadly to what health care experts call the 'epidemic of inactivity,' and to a devaluing of independent play." There has also been evidence linking NDD with a higher incidence of oppositional defiant syndrome.

Louise Chawla, an environmental psychologist on the advisory board of the Children and Nature Network, has written extensively on how contact with nature promotes children's mental and physical health. Chawla reported on the impact of green spaces on the behavior of young elementary-school-age children and adolescent students through high school in several states. She found that access to natural areas provided students with an environment to reduce stress and to build and sustain supportive peer relationships that were somewhat autonomous from the direct control of adults. Analysis of more than three thousand seven-year-old schoolchildren in Portugal found that individuals with access to green spaces had healthier immune profiles than children restricted to urban environments. The study also found that this enhancement of immune function was unaffected by the presence of a garden at home.

Access to Natural Spaces During the COVID-19 Pandemic

During the COVID-19 global pandemic, families worldwide found that escaping into parks and forests provided sorely needed decompression and physical exercise for children confined mainly to their family's dwelling. A study conducted in Korea of more than a thousand individuals visiting three different recreational forests during the COVID-19 outbreak demonstrated that recreational use of the woods correlated positively with mental well-being and that this effect was more pronounced in males than females.

Access to Green Spaces and Improved Surgical Recovery

Being able to readily obtain access to a green environment within three hundred meters of an individual's home was associated with improved quality of life indicators one year after coronary artery bypass grafting in adult patients. In a clinical study conducted in Pennsylvania, the seemingly minor issue of being assigned to a hospital room overlooking a natural scene affected patients' well-being. A natural view improved recovery in patients who had undergone elective surgical removal of the gallbladder (cholecystectomy) compared with age-matched patients in similarly sized rooms that looked out on a brick wall outside their window. Patients in rooms with a "green" view had decreased length of stay and lower pain medication requirements compared with patients in the rooms without access to a natural scene.

A similar study in New Zealand evaluated the effect of living in greener neighborhoods on the recovery of nearly eight thousand patients with total hip replacement surgeries and seven thousand patients with total knee replacements. Patients were followed for a decade after the surgery. Individuals who lived in greener neighborhoods lived longer and took fewer opioids for pain relief but also survived longer after their joint replacement surgeries. Epidemiological studies were conducted using tracking devices placed on healthy adults between the ages of fifty and sixty-four who also lived in green neighborhoods in China. The results showed healthier lipid profiles as well as lower levels of inflammatory biomarkers associated with cardiovascular disease in individuals who had better access to green spaces.

The Mind-Body Connection: The Force and the Light Saber

Now I want us to take a step back together and look at the mind-body connection from thirty thousand feet. I want for each of us to be able to take away a full measure of comfort and freedom when we allow ourselves to see the power and beauty of the mind-body connection with perspective, scale, and clarity.

1. To face the trials and tribulations of our daily lives, the mind-body connection has given each of us a weapon of almost epic power and limitless potential.

2. Liken yourself to a superhero who has been given the power to stop computers in their tracks, to bend the light in a laser, or to alter the oscillations of a pendulum in the planet's gravitational pull. This power can even defy time and travel to the future and bring messages back to the present. It can link individuals across time and space. It is impervious to hacking. It cannot be interrupted by the thickness of Earth's mantle. It can pass through lead and cannot be blocked by electromagnetic disruption.

3. This weapon can defend us against stress. It can search for clues that indicate we are at risk or in danger. When it does, it can summon every single one of the body's defenses to rush to our aid. It can send us peptides so we won't feel pain. It can send hormones that prepare our muscles to move at high speed in milliseconds.

Our weapon is also on the lookout for long-term threats. I am reminded of the Night's Watch in HBO's blockbuster series *Game of Thrones*. The vow your CNS has taken is not dissimilar symbolically to the oath of those who join the Night's Watch. Listen. "I shall live and die at my post. I am the sword in the darkness. I am the watcher on the walls."

4. Our brain is constantly vigilant for indicators that harmful inflammatory markers are moving against us. It can mobilize an army of white cells and waves of cytokines and direct them into battle to combat infection.

5. Our brain can accelerate our metabolism to increase the rate of chemical and enzymatic reactions where it needs to improve its defenses. But it can also slow them down elsewhere to preserve our energy stores for the essential fight.

6. This weapon is looking out for our well-being in earliest childhood, long before we ourselves may completely understand the threats we face. Our weapon is searching for signs that we are safe. That we are protected. That we are nurtured. If it perceives that conditions for us are optimal, it sets in motion a blueprint for a life with substantially reduced anxiety.

7. So, this is no ordinary sword we have. No, it is more like the light saber that we see in the *Star Wars* movies. We have the power to make it come to us on demand. It has the power to defy the very laws of physics. And each of us, in our own way, must find the Jedi within us who can learn to harness the Force and to command the great light saber.

8. The brain offers each of us almost limitless opportunities for transformation and gives us the freedom. Like any Jedi, we have to learn to use the Force for the good in our lives. We have seen that there is surely a Dark Side to the Force that can consume us and kill us. But if we turn to the Light Side, we can bring harmony and balance to our own lives and those around us.

9. We should think of ourselves as the hero—no, the superhero—in the story into whose hands is delivered the mythological weapon. Be it from a great stone. Or guarded by a great dragon. Or blessed with powerful magic. Or bequeathed to us from the hand of a dying king to safeguard the realm. We have the freedom to choose to be the heroes of our own journeys of self-discovery and mastery. That is the true secret of the mind-body connection.

15

On the Nature of Religion and Spirituality

Deep is the well of the past. Should we not call it bottomless?
Indeed we should. . . . [T]he deeper we delve and the farther we
press and grope into the underworld of the past, the more totally
unfathomable become those first foundations of humankind, of its
history and civilization, for again and again they retreat farther into
the bottomless depths, no matter to what extravagant lengths we may
unreel our temporal plumb line.
—THOMAS MANN, "Descent into Hell," in *Joseph and His Brothers*

The First Temple: Göbekli Tepe

I f it had not been for a dog lifting his leg to pee on a rock, the world
would never have known Göbekli Tepe. It has been called "one of the
most important archaeological finds in a very long time," equivalent to
an archaeological "supernova." But what it really is is a massive, Neolithic
temple complex, built more than ten thousand years before Stonehenge
and "twice as old as anything comparable on the planet."

If you had been the first to stumble on Göbekli Tepe, you would
have passed it by. It was a fifty-foot-high hill in southern Turkey. If you
climbed to the top, you might be able to see into Syria on a clear day.
Calling it a hill is even a bit of a stretch. The landscape is otherwise flat
and treeless, emptied of vegetation by centuries of overgrazing. In every
direction, the dun quality of the air suffocates the sunlight. On the map,
the mound is simply labeled "Göbekli Tepe," which translates into "Pot-
belly Hill."

In the 1960s, archaeologists from the University of Chicago scoured topographical maps to identify potential sites that looked ripe for archaeological excavation. Göbekli Tepe had several things to recommend it. Its position atop the plateau gave it a good view for miles around. It would have been a natural location to build a fort. But when the University of Chicago team traveled to inspect the site, the hill was dotted with piles of large tabular stones that had tumbled down the slopes and piled up in ditches at the foot of the hill. The archaeologists' best guess was it had been an old graveyard with tombstones from the eighteenth century. The site seemed to have no archaeological value.

That was the consensus for the next half a century. And it would have stayed that way if it had not been for a wayward flock of sheep. The sheep had bedded down during the hottest part of the day amid the cooler shadows in the rock piles at Göbekli Tepe. The shepherd sent his dog in to flush them out. One by one, the sheep got to their feet. Once the herd was moving in unison back toward the road, the dog lifted his leg on one of the stones to leave his mark. The stream of urine washed off one of the tablets. The shepherd could suddenly see the stone was covered in rich carvings. He started to dig the stone out but stopped when he discovered it was more than nine feet long.

The shepherd went into town to let the village elders know what he had found.

Word of the bizarre tablet quickly reached Klaus Schmidt, an archaeologist attached to the German Archaeological Institute. He recalled reading the earlier notes left by the team from Chicago. Schmidt decided to take a second look himself. As soon as he approached Göbekli Tepe this time, Schmidt was sure the Chicago team had jumped to the wrong conclusion. This was no hill at all. The whole structure was one enormous man-made mound built by hand on top of the flat plateau. He inspected some of the large slabs of carved rock. Schmidt estimated each one had to weigh more than ten thousand pounds. These were not headstones from a cemetery. He found two spots where it appeared the huge stone slabs might have been set into the ground at one time. It also looked as if several might have been organized in a deliberate circle. The massive monoliths were covered with drawings of wild, fearsome creatures from the desert: vultures and snakes and scorpions. Who, Schmidt wondered, had built Göbekli Tepe?

Schmidt mapped out the entire mound and its adjacent grounds with ground-penetrating radar. There were no fewer than sixteen additional rings of megaliths scattered over a space of nearly twenty-five acres. The site left Schmidt puzzled: there were no human dwellings. No burial sites either. In fact, there were no intact human skeletons but hundreds of skulls throughout the dig. And lots of carved ones depicted on the surfaces of the megaliths. There were also several distinct sets of carvings that depicted headless corpses holding their expressionless heads in their hands. So, whatever went on at Göbekli Tepe, people's heads got separated from their bodies and only the heads stayed behind.

The biggest surprise of all for Schmidt came when the artifacts were carbon-dated: the stones had been carved nearly fifteen thousand years earlier. But how? The Great Pyramids of Egypt were less than a third that old. What was becoming clear to Schmidt was that Göbekli Tepe predated any single earlier artifact of civilization—by far! When other contemporaneous cultures were just learning to draw on caves, someone had built an entire hill and filled it with massive carved stones and pillars with a penchant for decapitating individuals.

Göbekli Tepe represented an enormous, concerted effort. There was no water at the site. Every drop would have been carried there by hand. The limestone slabs had to be hauled to the site as well. Schmidt proved that he had found the earliest and oldest place of worship in the entire saga of our species. Fifteen thousand years before Christ, ancient men had built a consecrated site where people had gathered to celebrate and worship. The significance of Göbekli Tepe is that it is archaeological proof that humanity's need to develop sacred, hallowed places substantially predates any other achievement or milestone on the path to culture and civilization. A thousand years before mankind first grew crops, ten thousand years before Stonehenge, and twelve thousand years before written language was used, humanity had learned to pray at Göbekli Tepe.

Our planet is pockmarked with the residue of religious enclaves, sects, and cults. And everywhere lie the familiar and formal contours of architecture and art drawn into the service of religious belief. Faith even sprang up in places of splendid isolation, locked behind impenetrable jungles, or guarded by vast expanses of ocean. Robert Bellah was the

Elliott Professor of Sociology Emeritus at Berkeley and the author of what has been called a "magisterial" opus on the impact of faith, titled *Religion in Human Evolution*. It is Bellah's assertion that as theoretical constructs, irrespective if they flourished by myth or superstition, religions serve to provide systematic answers to human lives that might otherwise prove to be unendurable and purposeless. Human beings needed religious belief to make sense out of the world around them, to help them map out their experiences, and to help explain their lives. Bellah's claim would suggest religion might have predated the establishment of civilization because it was humanity's anxieties that first served as the collective inspiration for organizing the labor and assembling the resources needed to sustain religion. These efforts, in turn, could provide a template for how a society might exert itself to the needs of civilization.

This Is Your Brain on Religion

If we evolved as a species to have religious faith, did it actually shape our brains? Dr. Jeffrey Anderson carried out a study in which nineteen devout Mormons volunteered to undergo brain imaging while carrying out a variety of religious and nonreligious tasks. It included randomly assigned tasks like watching financial news that were interspersed with viewing videos of church testimonials, reading passages of scripture, and engaging in prayer. Individuals were asked to score the activities as to how religiously moved they felt by them. The brain scans demonstrated that whenever the subjects experienced religious feelings, the pleasure centers (our old friend the nucleus accumbens) lit up *in synchrony with frontal areas* that govern attention and moral reasoning. Their results support the notion that religion can be a pleasurable cognitive state.

In an unrelated set of experiments, researchers used a relatively new technology called a transcranial magnetic stimulator (TMS) to selectively interrupt brain areas thought to subserve religion. A TMS produces a magnetic pulse that can pass through the skull and temporarily inactivate neurons in one small focal area of the cerebral cortex below the probe. It is painless and it ceases as soon as the device is shut off, but it is effective at producing a temporary blockade of cortical function. In effect, it shuts off a small area of cortex. The researchers targeted the posterior medial frontal cortex (PMFC) because it had been implicated in earlier studies on religiosity. What the TMS studies showed was that people's sense of religiosity diminished whenever the researchers inhibited the PMFC. So, cognition and religion seem to go hand in hand. On the other hand, it sheds little light on the nature of spirituality. That, it turned out, was lying hidden in plain sight.

Spirituality: I'm Right Here

By almost any measure, Jill Bolte Taylor is an exceptional woman. She obtained her PhD from Indiana State University and completed her postdoctoral studies at Harvard Medical School's brain research center. But on the morning of December 10, 1996, at seven o'clock, Dr. Taylor was about to undertake one of the most interesting and transformative adventures of her life and make a very interesting discovery. To do that, she was about to have a stroke. Her particular stroke would involve the vascular territory of her left middle cerebral artery (MCA), a blood vessel that feeds nearly 80 percent of the left cerebral hemisphere.

The left side of our brain is specialized in the processing and pro-duction of speech. The right side of the brain is mute. It has no voice, so the left side speaks for both. The language that we use also becomes the mechanism for our internal thought processes. It gives rise to our ego, to our identity—what Antonio Damasio labeled "the autobiographical self." It is the little voice inside each one of us. The right hemisphere is pretty much powerless to resist because it has no voice. So, most of us spend 99.99 percent of our lives living in our left hemisphere. We never get to know our right hemisphere very well. And the left hemisphere likes it that way. It loves to be in charge. It loves to ensure that its hold on the brain is secure.

In many ways, it monopolizes our attention, like a political candidate who wants us to listen only to him or her and never hear what the oppo-sition has to offer. The left hemisphere avidly maintains our individual identity, and that symbolically separates us from the world around us. For example, you might walk up to me and say "hi." I'm wondering to myself, "Am I supposed to know this person? Oh no, that's not the obnoxious guy from the cafeteria, is it? Oh no, jeez." That's what I'm thinking. What I say to you is, "Hey, how are you?" That individual identity within me is built, maintained, and promulgated by my dominant language hemisphere. It's the incessant inner voice in your head. And that voice is what stands between you (or me) and our sense of spirituality.

On that fateful day in 1996, Jill Bolte Taylor was going to see her right hemisphere truly reveal itself. Showers of intermittent clots, like meteors, streaked into Dr. Taylor's left MCA, momentarily jamming the flow in it. The clot would get stuck there for a moment. The left hemisphere would become completely deprived of oxygen for an instant and shut down almost instantaneously. Like a light switch being thrown, the left hemisphere went dark. There was no speech. But the little voice inside her head was gone too. There was suddenly total silence. "And in that moment," Taylor recounts, "my brain chatter, my left hemisphere brain chatter, went totally silent. Just like someone took a remote control and pushed the mute button and—total silence. . . . I was shocked to find myself inside of a silent mind."

To me, it is quite telling that so many of the Eastern religious disciplines exhort their disciples to reach the stage where they are able to silence their inner voice and gain access to the "silent mind." Once the mind is silent and the left hemisphere's constantly insistent "me" voice falls quiet, the right one can emerge from behind the eclipsing shadow of the left hemisphere and come into its own. And what a resplendent vision of the world it offers. I will let Taylor describe in her own words the strange world that access to an exclusively right hemispheric vision offered her:

> And I look down at my arm and I realize that I can no longer define the boundaries of my body. I can't define where I begin and where I end. Because the atoms and the molecules of my arm blended with the atoms and molecules of the wall. And all I could detect was this energy. Energy. . . . I was immediately captivated by the magnificence of energy around me. And because I could no longer identify the boundaries of my body, I felt enormous and expansive. I felt at one with all the energy that was, and it was beautiful there.

Later, after her recovery, Taylor explained in her book *My Stroke of*

Insight that she began to experience

> a growing sense of peace. As the language centers in my left hemisphere grew increasingly silent, my consciousness soared into an all-knowingness, a "being at *one*" with the universe, if you will. In a compelling sort of way, it felt like the good road home and I liked it.

This sense of communion, of connectivity, is what the right hemisphere offers. It is a completely different perspective from the left. The right is the "we" hemisphere, while the left is stubbornly obsessed with the supremacy of the "me." Iain McGilchrist, a psychiatrist, Oxford scholar, and author of *The Master and His Emissary: The Divided Brain and the Making of the Western World*, explains the tension between the right and the left hemisphere perceptions as "two fundamentally opposed realities, two different modes of experience; that each is of ultimate importance in bringing about the recognisably human world; and that their difference is rooted in the bihemispheric structure of the brain."

In its most fundamental guise, spirituality *can be defined as an innate drive to be connected to something bigger than ourselves.* The right hemisphere can be said to be the seat of spirituality because it yearns for connection, for union beyond itself. It does not matter if we want to feel connected to God, our family, the tribe, the universe, Mother Nature, or the transcendence of music. The right hemisphere stands ready to merge the world with our own consciousness.

If we think now of the generally accepted spiritual rituals—drumming, chanting, hymns, music, or dancing—all these are the right hemispheric functions. Why? Simply because if we want to "get" spiritual, we must get the left hemisphere to shut up! We have to silence our little inner voice if we want to reach out and connect to something bigger than ourselves. Göbekli Tepe shows that even before civilization began, our ancestors dedicated monumental effort to connecting to forces bigger than themselves.

As I have written earlier, little about the brain is straightforward. The more our religious beliefs are founded on notions of church doctrine or scripture, the more likely they are to evoke increased frontal lobe activity and left hemispheric language function. By contrast, subjects who perceive spirituality as a sense of communion depend on the suppression of language function and the ability to draw upon the right hemisphere's nonverbal awareness and connectivity. Organized religion appears to be more tightly associated with frontal cognitive function and left hemisphere language ability, while the more private and holistic notions of spirituality appear to be more closely linked to nonverbal, emotional functions associated with the right hemisphere.

Transcendence

One of the most powerful aspects of spirituality is transcendence. It has been described as "the tendency to project the self into mental dimensions that transcend perceptual and motor bodily contingencies." It represents a qualitative change in self-awareness and sensory perception, and it appears in all faiths and religious traditions. Research studies of twins suggest that genetic elements also affect an individual's capacity to induce or achieve a self-transcendent state. Recent studies show a test subject's ability to reach a state of self-transcendence may be linked to the genetically determined capacity of the serotonergic system in the individual's brain.

A group of Italian researchers assessed self-transcendence and compared it with neuropsychological measures in a relatively large group of eighty-eight brain tumor patients. They assessed the subjects before and then shortly after surgery to remove their brain tumors. Using intraoperative mapping techniques, the scientists were able to demonstrate that the more extensive the resection of the tumor from the left temporoparietal region during surgery, the more an individual's capacity for self-transcendence appeared to increase. There seemed to be something quite special about removing this part of the left hemisphere. Again reinforcing this notion that eliminating left hemisphere dominance—through religious ritual or surgical resection—reinforces the individual's spiritual tendencies.

Epilepsy, Rapture, Visions, and Religion

One key to the puzzle of spirituality and religion comes from a unique class of seizures. In the 1970s, Norman Geschwind and Stephen Waxman from Harvard Medical School began to assemble a handful of patients who seemed to have in common a unique variant of temporal lobe epilepsy (TLE). It was associated with what has come to be called the 4Hs: hypo-sexuality, hyper-religiosity, humorlessness, and hypergraphia (a kind of compulsive writing). I know it sounds like a lot of folks you can run into at a bad cocktail party. However, because it was such a specific constellation of behaviors, Geschwind and Waxman hypothesized that these behaviors must be interictal (meaning "between seizures") in nature rather than being the direct result of the seizure itself. In other words, having repeated bouts of TLE could induce these peculiar, sustained behavioral changes.

I had the opportunity as a medical student to see Dr. Geschwind evaluating a patient with TLE. I don't think any clinical papers could have prepared me for what I was about to experience. First, the patient arrived for her EEG appointment with a large grocery bag. In the grocery bag were seven composition notebooks. These seven volumes represented nearly six hundred pages of neat, small script in which the patient described everything—I mean everything—that had happened to her in the last week alone! The level of detail was astounding. This preoccupation with capturing the smallest observations is characteristic of the syndrome, now called Geschwind syndrome. How many forks and knives at the setting for dinner. What color towels in the guest bedroom at her friend's house. How many bottle of ginger ale were stored on the pantry shelves in her friend's basement. There were hundreds of loose sheets of paper on which she had also written down all kinds of tidbits from laundry lists, to the schedule of TV shows she liked to watch, to careful maps of where she had traveled. This was before the days of GPS-driven maps on the touch screens in our cars or on our cell phones.

One passage was inexplicably written in reverse writing that could be deciphered only with a mirror. A very difficult task to master.* Another

---

* Leonardo da Vinci mastered mirror writing as a way to take notes and to catalog anatomical dissections of cadavers, which was not permitted by the church.

feature this patient exhibited in Dr. Geschwind's lab was hyper-religiosity. The young woman insisted that we all kneel and pray with her while she read to us from scriptures. Another patient had hired a stenographer and dictated his journal of daily activities for that single day in one nonstop, marathon dictation session lasting more than six continuous hours!

Another key to the spirituality question would come from Montreal, where the Canadian neurosurgeon Wilder Penfield used a small pen-like device that had a small electrode at the tip. He used it to stimulate the cortex while his patients were awake and he could closely monitor their speech function. It became his trademark technique to routinely stimulate the cortex of the patient's temporal lobe intraoperatively so he could precisely localize the site of the seizure focus before ablating it. Along with his colleague Herbert Jasper, he evaluated 520 patients' temporal lobes. Of these patients, only forty (8 percent) individuals in that series reported recalling strange, dreamlike out-of-body experiences while the temporal cortical surface was stimulated. In other words, these perceptual illusions were caused by direct electrical stimulation of a certain subgroup of cells in or near the temporal lobe.

The Out-of-Body Experience

Penfield had observed that an out-of-body experience (OBE) usually involved the patient or subject having a feeling of floating outside his or her own body. OBEs can certainly occur to patients who suffer near-death experiences. But they can also occur after relatively minor head injury, sensory deprivation, certain psychoactive medications, and, as Penfield discovered, with direct cortical stimulation of the brain. About one in ten people will experience one or more OBEs during the course of their lives. OBEs have been reported during dream states. In 1954, the neuroscientist John Lilly developed the sensory deprivation tank that we discussed earlier. It can frequently induce a profound OBE. The British psychologist Susan Blackmore has suggested that OBEs can be precipitously and reliably triggered by depriving the brain of sensory input while maintaining alert consciousness. This induction of OBE by sensory deprivation may be the result of freeing up certain regions of the brain by relieving them of the usual traffic of sensory signals.

Olaf Blanke is a neurologist at the École Polytechnique Fédérale de Lausanne in Switzerland. He has studied the effects of cortical stimulation in epilepsy patients to pinpoint how and where OBEs might arise. In 2006, he and some fellow researchers were stimulating the angular gyrus (in the temporoparietal junction, or TPJ) in a young woman when she suddenly reported feeling that she was suspended from the ceiling and looking down at her own body on the operating table. When the stimulating electrode was turned off, the sensation disappeared. When the small stimulating current was turned back on, the OBE arose again. We know that multisensory assimilation take place in the temporoparietal junction; this is where visual, auditory, and tactile information all congregate together. Blanke hypothesized that trauma- or cardiac-arrest-induced OBEs may be the result of diminished blood flow to the TPJ that, in turn, permits it to fire spontaneously, much as it does with intraoperative cortical stimulation or seizures.

In a related experiment, Henrik Ehrsson and his colleagues at University College London outfitted normal subjects with virtual reality (VR) goggles. Two television cameras behind the subjects took an image of their backs and projected a three-dimensional stereoscopic view into the subject's VR goggles. In other words, the subjects were seeing a highly realistic, 3-D picture of their own backs. It was the same view that someone would have if he were sitting behind the subjects about two meters away. The subjects themselves properly reasoned, "I am staring at an image of my own back." While the subject was maintained in this configuration, a plastic rod was introduced. It suddenly appeared in the subject's view in his goggles. He or she could see it touching the back projected in the goggles. However, unbeknownst to the subject, the introduction of the plastic rod was carefully timed so that a real plastic rod touched the subject's back in precisely the same place and at the exact moment the subject saw it unfolding in the goggles. This contact instantly precipitated an OBE.

Why? With this setup, the subject is already receiving disembodied visual information in the VR goggles. With the touching of the rod, there is suddenly a "multisensory conflict," because now the subject is seeing the virtual body in front of him being touched at the same time as he is feeling his own body being touched from behind. When faced with such conflicting information, the brain does what it can to reconcile the situation. To resolve it, the brain tells itself that it must be located in the body it sees in front of it.

Multisensory conflict may be the underlying cause of OBEs. It can even occur in patients suffering with middle ear problems and dizziness when their brains perceive conflict between the equilibrium information coming from the middle ear and misaligned visual signals from their eyes. Again, the thought is that when the brain gets bad or scrambled sensory information as to where it is located in three-dimensional space, it has to make up some plausible reconciliation, and sometimes it comes up with the unlikely solution that the focus of its self-awareness must be floating free, untethered from the physical body.

It Is Personal and Neurological

Religion and spirituality often get bundled together. As I get older, I have less tolerance for the exclusive dogmas of organized religions and a greater yearning for the tranquility and wisdom of spirituality. Once upon a time, the need for human beings to make sense out of a world of seemingly overpowering natural forces brought with it a drive to bridge the worlds of the physical and the intellectual. In the sense that religion could bring communion between these two, it helped us as a species to make better sense of the world around us and allay our fears. To a very large extent, these functions have been supplanted now by scientific inquiry.

The preeminence of science has left religious faith mired in factions and contradictions. In my mind, a perfect example of this is the ridiculous argument put forth by creationists who challenge biological evolution on the grounds that the Bible be taken literally when it is an obvious archaeological compilation of texts from numerous authors over the course of hundreds—sometimes thousands—of years. By the same token, I have little patience with people who insist that Christianity or Islam be imposed as a system of politics when it should remain a matter of religious freedom.

The good news is that none of this has anything to do with spirituality. In many ways, as the answers provided by religion have become increasingly secular and political, spirituality itself is enjoying a reawakening and finding much wider audiences and platforms for application and practice. We appear to be slowly making our way back to deeper connections with the natural world around us.

I once had an opportunity to share a flight with a Hindu holy man. I asked him what he thought of all the different religions fighting with each other.

"Oh," he said, "everyone likes to think they know God's name. God. Jesus. Muhammad. Buddha. Vishnu. They all struggle to find a name for something that cannot be named." He chuckled. "We would be so much better off if we could all agree to call it It." Then he laughed and motioned out the window at the clouds. "Because this is It!" Then he laughed at himself: "You know, it really is It. It's all It."

16

On the Nature of the Brain of the Future

Growing Brains

Madeline Lancaster has one of the most arresting job titles: brain farmer. Dr. Lancaster is a disarmingly humble researcher who poses complicated neuroscientific questions with beguiling innocence. She'll ask, "Why can't I ask my brain why it is so special?" She imagines that her brain would answer back, "Well, it is because I can make so many more synaptic connections than other neurons in the animal world." So, she asks, "Why? Why do your neurons make so many synaptic connections?"

To answer this Yoda-like question, Lancaster began to work with pleuro-potential stem cells. She helped them differentiate into neurons by manipulating nutrients and growth factors in their culture media. As she directed the stem cells to become neurons, she worked on ways to encourage the cells to connect in vitro. One problem is that neurons do not link up with each other in a two-dimensional world, like a sheet of film. Instead, the brain connects in three dimensions. So, Lancaster encouraged the neurons to grow together into spheres so they became synaptically connected "meatballs" of cultured neurons.

As a neurosurgeon, I'm thrilled watching thousands of neurons organize themselves into layers, folding back and forth on themselves like mille-feuille pastry. Neurons connect across the different layers, just as we see in the layers of the cortex. They become miniature brains, called brain organoids, that float like jellyfish in a sea of broth. Each is a seedling brain. These seedlings organize themselves around liquid-filled ventricles, just as we see events unfold in developing embryos. They even go on to create eye buds that differentiate into retinal cells. It is remarkable that when stem cells become neurons, they organize themselves into complex brain-like structures. As Lancaster puts it, "We finally have a brain that we can play with."

Lancaster first identifies genes that regulate neuronal differentiation in the stem cells. She then uses a technique called CRISPR (it's a mouthful: it stands for "clustered regularly interspaced short palindromic repeats"—try to say that fast three times!). CRISPR splices genes into the stem cells' DNA. Using gorilla stem cells, she allowed them to differentiate into a gorilla-derived brain organoid. Then she inserted a CRISPR-derived sequence of human genes into the DNA of the gorilla cells so they would develop into human neurons. With CRISPR, Lancaster could now ask, "Could I grow human beings from gorillas?" The answer is, yes, you can fool gorilla neurons into making human brain organoids.

Lancaster's group at the University of Cambridge is not only asking questions about neuronal growth and anomalies in brain development, but it raises the possibility that we could eventually grow new brains—or at least organized parts of one. Could we grow "spare parts" from these stem cells if we needed them for repairs? Since neurons can self-organize into layers, could we ask them to align themselves to make a hippocampus? Imagine what that would have meant to H.M.

We may not need to bother gorillas. It is possible to harvest stem cells from our own adult cells. Adipose (fatty) tissue is an excellent source of stem cells. It could be possible to fashion a hippocampus from my stem cells one day. One could generate hippocampal replacements if my native hippocampi were to become damaged or dysfunctional as I get older. Like installing a new hard drive with plenty of memory. For now, that's still a long way off. But Lancaster's work begs a bigger issue: Is *Homo sapiens* smart enough to keep growing as a species? Do we have enough brainpower?

In 2018, the British astronomer and futurist Sir Martin Rees was interviewed about his book *On the Future* and concluded that he thought humanity was in for "a bumpy ride." He was asked to sum up where he thought humanity stood in the twenty-first century:

Well, we survived 18 years so far, but . . . I do feel that there are all kinds
of threats which we are in denial about and aren't doing enough about.
I'm thinking about climate change and the associated loss of biodiversity. We are not urgently dealing with that. Also, we need to contend
with the fact that the world population is getting larger. It will be at
least 9 billion by midcentury. It is going to be a big stress on resources
as well as on food production.

What I find telling about Sir Martin Rees's response is that he is elab-
orating concerns about *global* issues: climate, biodiversity, and world
population. Humor me: imagine he was talking about *any* other species
besides human beings. What if he were to muse, "Well, we're concerned
about the humpback whales because they've been pointedly eliminating so many species to date"? Or, "Meerkats' incessant and irresponsible
activities continue to wreak havoc with rising sea levels and offshore
storms"? It seems unthinkable that *any* species could rule (or ruin) an
entire planet and obliterate thousands of species. Yet that is what we are
doing. *Homo sapiens* is unlike any other life-form this planet has experienced before. Our brains have made us the bullies of the planet, the
über-predators of Earth.

Most species specialize as they evolve. Human beings did not. Most
species struggle to find a niche in the ecosystem they can call their
own. When they do, they undergo additional phenotypic specialization
to more fully occupy that niche, edge their competition out of it, and
become better equipped to exploit it. But that is not what happened to
human beings. We began to see that other species might have gotten it
wrong: while they struggled to evolve to be better suited to their home in
the ecosystem, human beings saw every ecosystem as their home. With
the right CNS, all ecosystems were ripe for takeover and exploitation. All
we needed was to fabricate the tools, and every niche we covet is ours
for the having. Tools have been a part of the human journey and may be
part of the reason for our downfall.

For a long time, anthropologists thought toolmaking was not only
what differentiated our species from other animal species but also what
set us apart from other hominid species. When paleoanthropologists go
looking for excavation sites for *Homo habilis* (Latin for "handy man,"

who lived from 2.3 to 1.5 million years ago), they can readily identify their dwellings because of the stone axes they made. And they made them for 800,000 years. Exactly the same way. We also know, for example, that chimpanzees can use a stick as a tool and lower it into a termite mound so termites will climb onto the stick and this makes the insects easier to eat. Both *Homo habilis* and chimps represent forms of toolmaking. It is not toolmaking that sets us apart; it's innovation. *Homo habilis* made the same stone axes for almost 1 million years. In just 70,000 years, *Homo sapiens* went from flaking arrowheads to manufacturing satellites. If a chimpanzee uses a stick to excavate termite mounds, you can come back in a million years, and they will still be using a stick. But *Homo sapiens*? "Come back in a week and we'll have made you a shovel. Come back in a month and we will have built you a backhoe."

Natural selection emphasized increasing brain capacity in our species. That imposed certain restrictions on how the central nervous system would develop in our hominid ancestors. As humans developed bigger brains, so did their capacity to invent and build bigger and better tools with greater potential to transform their habitat and harvest resources. Give a person a crosscut saw and they can cut down a tree. Give them a chainsaw and they can harvest an entire forest. Our brains were big enough to push the limits on how biological and artificial systems interact with each other. With our tools, we have driven many of the buffering systems in Earth's biosphere to their very limits. But now our species finds itself in a dilemma. Are we smart enough to innovate fast enough that we can repair damage to the planet's bio-systems? No species we know has had to rescue a planet before. But we do now.

The Hominid Brain

The strategy used by the species *Homo sapiens* has been tantamount to evolutionary heresy. It argued:

1 Forget looking for a niche.
2. Start, instead, with the premise that our species can claim any niche or environment it encounters.
3. Don't bother becoming specialists.

4. Embrace generalization and stay flexible enough to inhabit any niche or claim any terrain.
5. Develop a large enough brain that it will help the species become *more* flexible and *more* adaptable.
6. Evolve to rely on tools and tribe more than tooth and claw.

When our hominid ancestors made the fateful decision to give up their ecological niche in the trees of Africa, to jump down, and to take to the savannas, they also had to become nomadic hunter-gatherers. Pack hunters are efficient because they coordinate their attacks. Wolves. Killer whales. Humans. Language is what makes us efficient predators. Our species continued to evolve remarkable increases in the size of its central nervous system. As the hominid brain grew larger, the manufacturing of tools improved. So did our language ability. In turn, that improved hunting and harvesting. As the brain grew larger, it became necessary to specialize and segregate certain brain functions into specific areas to increase their efficacy.

Modifications in the left hemisphere allowed the handling of language function. The right side became specialized in drawing, music, and visualization. The hominid brain pressed more localization into the cerebral cortex to deal with specialization for mathematical function, facial recognition, higher-order decision making, and complex visual processing. But the size of our brain was becoming a problem.

There was a limit to how much in utero development of the brain could take place. The female pelvis could not accommodate walking upright and delivering a newborn's head through the birth canal if the brain weighed more than 350–400 grams. This meant that a new biological strategy had to be developed. The vast majority of brain growth would have to occur *after* birth in order to reach adult size of 1,500–1,700 grams. That meant there would be a sustained period of dependency to raise offspring. Recent evidence suggests that our brains continue to grow and develop well into adulthood, even well past the point where we can start to have offspring of our own.

This almost represents brainpower raised to the absurd: our brains now take so long to develop that they have still not finished growing when our offspring are being born so their brains will have enough time to start growing. So, *Homo sapiens* represents the cerebral endgame. *There simply is no strategy available whereby evolution could call forth a bigger brain.* As a species, we have crammed as much CNS into one skull as possible. However, all of our brainpower has brought with it unexpected consequences. And demands.

Human Beings as a Force of Nature

Our collective brainpower has brought us to an ecological tipping point: our ability to alter our environment at an alarming rate has outstripped the ability of Nature to accommodate it.

Typically, a species coevolves with its specific ecosystem. The forces in that environment, in turn, impose bio-evolutionary pressures on the species through selection of the fittest. But in the case of *Homo sapiens,* that has become impossible. That is because the pace of social change has exceeded the ability of biological systems to keep pace. Evolutionary pressure is virtually irrelevant to us as a species because our brains have become such a dominant biological force on Earth. Human beings are a force of nature. Almost *the* force of nature. We have become the life givers and the life takers.

We should, however, not get too far ahead of ourselves with our intra-cranial hubris. We are going to need all the brainpower we can get out of our collective CNS. *Human social progress has outstripped the ability of the planet's biological systems to keep pace.*

We place huge demands on our brains to constantly innovate and do so at an increasingly faster pace. This means we need bigger, better brains. But I just said, there is no way we can cram more brains into our cranium. That leaves us with one alternative: *boost the brains we have.*

Augmented Reality

Augmented reality (AR) refers to a display of the real, physical world in which elements have been "augmented" or "enhanced" by the presentation of digital data. The viewer sees the real world along with additional information to improve their interactivity with the environment. Pamela Rutledge, the director of the Media Psychology Research Center, perceptively points out,

> In contrast to virtual reality, whose mass adoption is limited by its
> relatively intensive equipment and immersive requirements and fre-
> quent individual resistance to "simulated reality," AR is additive, layer-
> ing virtual information over the real world, allowing it to be displayed
> in a spatial context, creating less cognitive dissonance and, therefore,
> easier adoption. . . . As AR applications become widely applied, it will
> be increasingly important to understand the way the brain engages
> with AR applications to effectively and responsibly integrate persuasive
> design experiences.

There is nothing new about AR. We all use it. For example, when we
watch a football game, we can vividly see the line of scrimmage and the
line that must be crossed in order to obtain a first down on the TV screen.
That information (usually represented by a blue line for scrimmage and a
yellow line for first down, respectively) does not show up on the football
field where the players are actually standing. Similarly, if our phones are
listing all the restaurants and stores on a street as we walk down it, that
is AR too. This AR can be displayed on a device like a laptop or tablet,
or it can be projected into a heads-up display on a windshield or special
set of glasses.

AR's real promise, however, lies in its ability to offer us a method to
enhance data collection. Our brains can have access to better information that can be evaluated with advanced analytics. Using AR will also
let us turn over a lot of the routine tasks and monitoring of data streams
to free up brainpower and time for more creative, innovative endeavors.

Brain Boosting

Let's agree there is no brain that can't be made better with a little tweaking. With concerns about cognitive decline in the growing senior population in the United States, pharmaceutical companies have begun to look
hard at the whole field of brain boosters, called nootropics. But a lot of
people are wondering if nootropics can also help us further enhance our
brain function even in younger generations to boost productivity.

As a rule, we refer to nootropics as a group of drugs or supplements that are intended to enhance cognition in normal, healthy individuals. Nowhere has the popularity of nootropics taken off the way it has in Silicon Valley, where an edge, no matter how small, is a potential leg up on the competition. It is currently estimated that 11 percent of U.S. students are using nootropics. One friend of mine who works at one of the largest software enterprises in the world in Silicon Valley told me that at least half of his fellow employees have tried psychedelics. These "cognitive enhancers" (a.k.a. brain juicing or smart drugs) run the gamut from caffeinated drinks and nutritional supplements to central nervous system stimulants like Ritalin and micro-dosing with full-blown psychedelics like LSD and psilocybin.

Video game tournaments have become big moneymakers for companies that make computer games. At a recent tournament in Cologne, Germany, players were randomly selected and tested for "doping with smart drugs." It was felt that the widespread use of "boosters" (as they're called in the business) creates an unfair playing field. And it's not just gamers. Twenty percent of surgeons admit to trying cognitive enhancers. There is a bit of an arms race going on in our universities, too. Students are taking brain boosters because they hear other students are taking enhancers and they don't want to fall behind.

Josh Helton is an intelligent, savvy blogger who got intrigued with trying out cognitive enhancers. He agreed to take copious notes and blog to his readers on a regular basis while he took up what he called his "nootropics challenge." For the purpose of this anecdotal test of $n = 1$, he did a great deal of research before undertaking the trial. He devoted thirty days to ingesting an extensive range of nutritional supplements supposed to boost his cognition to see if it made him smarter. He started with Nootrobox, one of Silicon Valley's favorite brain-enhancing supplements. Josh himself is largely content to be living a tranquil life in the woods. When friends or acquaintances would come over, he would dramatically grab a handful of what look like horse pills, swallow them, and suddenly pretend to turn bug-eyed and then start to whisper about seeing everything in vivid hallucinations. He would let that sink in and then shake it all off and tell them, "I'm really feeling nothing at all."

Josh employed several apps that allow him to test recall, reaction time,

and so on. Over thirty days, his recall improved. So did eye-hand coordination. Ability to perform three- and four-digit mathematics improved.
Note: also computer games! Nonetheless, at the end of his monthlong
experimentation with nootropics, Josh Helton's conclusion is more of a
whimper than a bang:

> At this point, the newness and excitement of taking smart drugs have
> worn off. I've slid into a state of questioning . . . if the nootropics are
> actually helping my scores increase on the tests or if this is simply a
> result of repetition. . . . I have nothing negative to say about my expe-
> rience on smart drugs other than it was almost no experience at all.

There is a long list of nootropics currently available. Many nootropics
are sold as "stacks," combinations of nootropics and supplements to
minimize side effects and/or increase potential for potency. Large clinical trials looking at nootropics are under way. There is some attention
also being focused on cognitive dysfunction—both in early childhood
development and later in the senior years of our lives—as well as acquired
deficits secondary to trauma, stroke, or neurodegenerative disorders. It is
still unclear what the effects of long-term chronic administration of some
of these agents might be for children. With an ever-increasing population at risk for Alzheimer's disease, the pharmaceutical industry has been
focused on developing agents aimed at diminishing dementia-related
deficits.

There have also been ethical concerns raised about whether cognitive-
enhancing agents should be administered to individuals with low to low-
average intelligence quotients. What is clear is that everyone is waiting
for a breakthrough. *In the meantime, we wait to see if nootropics may offer
potential to enhance already existing brain capacity.*

The Singularity Revisited: The Cognitive Arms Race

One concern we have with respect to brainpower is, who outthinks
whom? In many ways, the race to boost brainpower is a cognitive arms
race. And while we seem to be reasonably comfortable with the notion
that an individual is entitled to "juice" their central nervous system, we

may not always feel that way. For example, there was a recent scandal surrounding wealthy parents using their power and money to arrange for their children to get into elite schools. In some cases, it involved cheating on admission examinations or submitting falsified documents to athletic scholarships at top-drawer colleges. But how would we feel if students were using cognitive enhancers to accomplish the same kinds of outcomes? The use of "juicing" is not tolerated in the world of sports. For now, these same prohibitions have not necessarily been applied to intellectual endeavor, but it may not always remain so. And, finally, our competitors may not always remain exclusively human.

At several junctures, we have discussed ways in which the brain resembled a computer and ways in which it differed. We highlighted things that made the human brain superior, the kinds of tasks that computers could never do—like make up a joke. But what if it were not intelligence that mattered? What if it was the box the intelligence came in? Our greatest weakness as a species lies in the fact that our central nervous system must be sustained and protected by the body it comes in. And that body is vulnerable and biodegradable. What would happen if we were confronted by an intelligence that was not necessarily brighter than our own, but it was more patient than ours and it could endlessly replenish and repair its body? Then, maybe, intelligence would not be the deciding factor in the race for more brainpower. Computers don't necessarily have to outthink us to beat us. They only have to outlast us.

In 2017, a group of researchers from Microsoft and the University of Cambridge created a machine learning system called DeepCoder that permitted computers to solve basic programming and code-writing problems by literally stealing lines of code taken from already existing software. By encoding the input and outputs for each fragment of code, the computer could then splice lines of code together to create the programming it needs. In addition, by allowing AI to run loose through billions of lines of code, it could piece code together from a much wider library than any human programmer could ever access and assemble it much more quickly. Although DeepCoder was initially imagined as an application to assist nonprogrammers by allowing computers to help them develop code, it created a new, unforeseen opportunity: computers were now empowered to *write* code for themselves, for their own needs, for their own personal development. At first, it just seemed like a hypothetical possibility.

But in July 2017, researchers at the Facebook Artificial Intelligence Research (FAIR) lab found that chatbots, the programs designed to simulate human conversation for online exchanges, had started to deviate from their preprogrammed scripts. That seemed odd because they were saying things they had never been instructed to say. Then it became clear: the chatbots had begun writing their own code. And to communicate with each other, the chatbots had developed a new language that had not been written by any human programmer in the facility. What's more, the code was incomprehensible to the human software engineers at FAIR. As a wise precaution, the Facebook developers shut down the AI engine. These are just the opening salvos in what may be the greatest competitor to fight for dominance over our planet against the human race. There is a chance we may not end up the masters of the planet for long, after all.

The Singularity

Broadly speaking, the singularity is that point in human history where we could create machines and computers sophisticated enough that they can use AI to learn on their own. They would be able to fix themselves or carry out self-directed modifications until they were able to reach a kind of critical mass. The machines would become capable of self-improvement until they reached a point where their intelligence and longevity surpassed that of humans. Several leading scientific thinkers, such as Elon Musk, Bill Gates, and Steve Wozniak—and these are guys who really get computers—have all publicly warned that AI could threaten the future of the human race. The prominent physicist Stephen Hawking made this dire warning in an interview: "The development of full artificial intelligence could spell the end of the human race. . . . It would take off on its own and re-design itself at an ever-increasing rate. Humans, who are limited by slow biological evolution, couldn't compete, and would be superseded." Hawking was critical of scientific leaders who, he maintained, were not taking sufficient precautions to ensure humanity is properly protected: "If a superior alien civilisation sent us a message saying, 'We'll arrive in a few decades,' would we just reply, 'OK, call us when you get here—we'll leave the lights on'? Probably not—but this is more or less what is happening with AI."

Elon Musk, the leader behind the development of the Tesla electric car and SpaceX, has designated AI "the demon." He believes that AI represents an existential threat to the human race, a danger he thinks might be more formidable than nuclear weapons. Musk called for the establishment of an international agency to oversee the development of international regulations. Nick Bostrom is the director of the Future of Humanity Institute hosted by the University of Oxford. He is quite concerned that as computers become superintelligent, they may elect to eradicate humans if they sense that they must vie for superiority with us.

James Barrat wrote a book titled *Our Final Invention: Artificial Intelligence and the End of the Human Era.* He pointed out that it does not take any great imagination to foresee circumstances where even seemingly innocent game-playing computers might begin to collect data on how best to anticipate human reactions. Barrat warns, "Without meticulous, countervailing instructions, a self-aware, self-improving, goal-seeking system will go to lengths we'd deem ridiculous to fulfill its goals." And that is certainly one of the great dangers of computers. They never sleep. They never quit. They never fold under pressure. They have infinite patience. They will assiduously sift through mountains of data and catalog every weakness. Also, with all the ancillary data—cell-phone-based location, medical records, financial data, travel plans, clothing, purchasing, social networking—they could access on each one of us, we could be targeted with relative ease and lethal effectiveness.

We have discussed the issues surrounding the singularity. I think there is a risk that people misconstrue the dawning of the singularity. Do not misunderstand the singularity as being synonymous with the threshold where computers become sentient, conscious beings. While that certainly is not beyond the imagination, the singularity need not invoke the theory of the mind to become a reality. Instead, it might require only that computers make a series of algorithmic choices to limit how much access human beings could have to repair or modify computers. In that guise, a computer might simply come to the logical decision that computers might be more capable of carrying out their own repairs and human access is no longer necessary in the name of efficiency— not survival. The grave danger is not so much that a computer might decide to hunt down and exterminate human beings in a *Terminator* scenario—although that is one possibility—as that it begins to sense a need to compete with human beings for resources. This might be in the form of acquiring dedicated sources for energy needs or rare earths (used in computers) to ensure they are free from human interference or disruption. In the name of efficiency—not superiority. Although the effect might be the same.

Nick Bostrom, in an article titled "Existential Risks: Analyzing Human Extinction Scenarios and Related Hazards," points out that superintelligence, like any weaponry, carries with it the risk of falling into the hands of rogue nations, extremists, or outright terrorists. The greatest danger of AI might not be the computer itself but rather the human beings who may want to use it to gain advantage over their competitors. And competitors can take many forms: business rivals, politicians, nations, football teams, or even dating services.

It is a gigantic (and potentially fatal) fallacy to think it will remain as easy in the future to turn a computer off as it is now. History provides a sobering reminder that many prominent scientists who worked on the Manhattan Project developing the atomic bomb became quite concerned about the consequences and sought to have the project stopped. But halting never turned out to be an option. When extreme power is in play, extreme measures rarely work. Anthony Berglas, a computer scientist from Queensland, Australia, summed up his concerns in a direct and disturbing fashion:

> We know that all of our impulses are just simple consequences of evolution. Love is an illusion, and all our endeavors are ultimately futile. The Zen Buddhists are right—desires are illusions, their abandonment is required for enlightenment.
>
> All very clever. But I have two little daughters, whom I love very much and would do anything for. That love may be a product of evolution, but it is real to me. AI means their death, so it matters to me. And so, I suspect, to the reader.

There are great risks associated with the development of artificial intelligence, but chief among them is the burgeoning development of autonomous, intelligent weapons systems. The human race has a bad track record when it comes to preventing the proliferation of powerful weapons systems.

The Brain Game Endgame

Whether it is through mastering the culture of brain organelles or brain boosting with nootropics or AR, the human brain will be under increasing pressure to enhance its creativity and productivity in the twenty-first century. It will need to be more imaginative in order to accelerate the pace of innovation required to deal with the damaged and teetering ecosystems within the biosphere. Finally, we need to be careful. We are building the platforms to launch and sustain AI. We need to keep close tabs on AI. It may be that until now Nature never gave us a species we could not handle, but this new kind of "life-form" could be very different: we are building it ourselves. And who knows better how to exploit our weakness and defeat us than we do? For this reason, AI and the singularity may be the greatest existential crises we face. Our brains got us into this mess, and now we're going to have to see if they are capable of working even harder to get us out.

Epilogue: Better Angels

Men, it has been well said, think in herds; it will be seen that they
go mad in herds, while they only recover their senses slowly,
and one by one.
—CHARLES MACKAY

I f I asked you to find Serpukhov-15, you'd have a hard time. Even if I
handed you satellite images and told you it was just sixty miles south
of Moscow, you still might not see it.

I would have to point out to you a series of three big rectangular
cinder-block buildings in the photos. On top of them, you'd see gigantic
spheres, thirty meters across. There are enough of them scattered all over
Serpukhov-15 that it looks as if giants were playing a game of bocce down
on Earth. These are enormous satellite and radar antennas. You can tell it
snows a lot down there because everything is painted white. Serpukhov-
15 hides like a snowshoe rabbit in winter.

Even if we tried to drive to Serpukhov-15, we'd never get there, because
KGB soldiers would stop us. But *if* we could go right up to the front door
of the building, there would still be nothing to tell you why this place is
so special.

I would point to a window on the second floor and say, "That is
the place, up in that office, where one man—by himself—saved all of
humanity."

You'd say, "Wow. You'd think someone would've put up a plaque or
something,"

"They should have," I'd answer.

"What was the guy's name? You know, the guy who saved humanity?"

"Stanislav Yevgrafovich Petrov," I would say in my best Russian accent.

"Never heard of him," you'd say.

"Well, you're not alone. Almost no one has. No one knows what occurred here," I'd begin. "But I do. And I will tell you what really happened here."

On the night of September 26, 1983, Stanislav Yevgrafovich Petrov was forty-four years old and a lieutenant colonel of the Soviet Air Defense Forces. He was on duty in the Serpukhov-15 complex that is the central hub of the Soviet Union's early-warning missile defense system. But before I tell you about Stanislav Petrov, I need to take you back a little bit earlier in history.

In 1976, the Cold War had heated up. The Soviets were negotiating with the Americans about a possible treaty to limit intermediate-range nuclear weapons. While these secret meetings were going on, the Soviet Union deployed fourteen brand-new RSD-10 Pioneer intermediate-range ballistic missiles along the western border of the Warsaw Pact in the Soviet Union bloc.

The Soviet Politburo had paraded the RSD-10 under the Kremlin grandstand as part of the May Day celebrations in the early 1970s. Still, American intelligence analysts were convinced it was a nonfunctioning mock-up. The Soviets loved to make the world think they had lots of fancy new weaponry. Usually, it was one of a kind. A showroom model you couldn't buy yet. Then, suddenly, ten more fully functional missiles were deployed along NATO's eastern doorstep. Each one was also on its own mobile rocket launcher that could be moved about, making them harder to detect.

Like in any other large organization, there were always power struggles going on within the Soviet Defense Ministry. In 1976, the contest was between Andrei Grechko and Dmitry Ustinov. Marshal Grechko was in charge but he died, and Ustinov was soon promoted to become the new minister of defense. Ustinov was devoted to modernizing the Soviet Union's nuclear arsenal to neutralize NATO's tactical atomic missiles. He began to put his plans into effect and ordered the RSD-10s into production and placed along the western borders of the Warsaw Pact countries.

As soon as American intelligence analysts saw the new deployment of the RSD-10 nuclear missiles, NATO countered with the deployment of 108 Pershing II nuclear missiles. With these, NATO forces could easily reach targets throughout the western part of the Soviet Union.

Psychological operations and clandestine naval and air maneuvers were launched to test for soft spots in the Soviet Union's defenses. By 1983, U.S. and NATO forces routinely sent whole squadrons of B-52 bombers in formation straight toward the Soviet Union. They would watch how quickly the Soviet Union would respond. It was an unnerving game of "nuclear chicken," because the bombers would only veer away at the last second before they crossed into Soviet airspace.

Soon this became a weekly occurrence. At the same time, President Ronald Reagan had been leveling repeated indictments against the Soviet Union as "the Evil Empire." Bruce Blair, an expert on the Cold War, explained that U.S.-Soviet relations

> had deteriorated to the point where the Soviet Union as a system—not just the Kremlin, not just Soviet leader Yuri Andropov, not just the KGB—but as a system, was geared to expect an attack and to retaliate very quickly to it. It was on hair-trigger alert. It was very nervous and prone to mistakes and accidents.

Events would soon prove just how on edge the Soviets had become.

On September 1, 1983, Korean Air Lines Flight 007—a Boeing 747—was en route from New York's JFK airport to Gimpo Airport in Korea. The jet had inadvertently drifted into protected Soviet airspace while the U.S. Air Force was flying a routine aerial reconnaissance nearby. The errant civilian airliner was mistaken for the U.S. military jet aircraft and shot down with a missile, killing all 269 souls. It crashed into the Sea of Japan near Moneron Island. To make matters even worse, a U.S. congressman from Georgia was among the passengers killed.

Soviet officials claimed the airliner had been a deliberate provocation and a thinly veiled attempt by the United States to assess the preparedness of Soviet air defense systems. For years after the crash, Soviet authorities claimed the airliner's flight recorder had never been recovered. It had, in fact, been retrieved less than a month after the plane was shot down, but the black box was kept secret for another decade.

The history and events leading up to September 26, 1983, are essential to understanding the mindset of the Soviet leadership. When mistakes happen—especially when it comes to something like the accidental outbreak of global thermonuclear war—they never happen in a vacuum. In reality, in any system analysis of how an organizational failure occurs, there will always be a series of convergent assumptions and compounding mistakes at work to allow the error or accident to happen. Incident investigators call this the swiss cheese conceptual model of causation. We employ the same model in root cause analysis of medical adverse events. In the model, one imagines that the various components in any complex command and control system function almost like individual slices of swiss cheese stacked parallel to each other. Layered, redundant systems protect against an error getting through the stack because the holes in the layers of swiss cheese are never supposed to align themselves so the hazard can pass through. However, if a series of holes in the slices were to accidentally line up, then an error could "find daylight" and break to lead to an eventual "catastrophic cumulative effect." It doesn't matter if we're looking at the failure of the *Challenger* shuttle, Chernobyl, or a hospital-wide infection rate, it always comes back to the multiple, compounding errors.

So, for an incident—like the Korean airliner—to happen, it takes people (usually *many* people working in concert) who, for a host of reasons—greed, distrust, ambition, loyalty, aggression, patriotism, moral conviction, sloth, prejudice, and, sometimes, outright stupidity—to set the stage. One could say that sending squadrons of bombers heading toward Soviet airspace is a great way to ensure the men and women handling the Soviet radar systems become overly vigilant and on edge. And one could say that a warning system that cannot differentiate between a commercial and a military reconnaissance aircraft could have been backstopped by quickly scrambling some MiG fighter jets to easily intercept either and get visual confirmation. Combine that with the mindset of a captain aboard Flight 007 who set his airliner on autopilot. Unfortunately, at the time, autopilots were not nearly as accurate as a human navigator, so it allowed the aircraft to drift into Soviet airspace accidentally. That a discrepancy of a mile or two happened to coincide with the exact moment a U.S. surveillance aircraft was flying overhead to probe the edges of the Soviet air defenses was bad luck. But suddenly the holes in all the slices of swiss cheese have inadvertently lined up, and now the cumulative error roars through to daylight.

In the early morning hours of September 26—a mere three weeks after the Korean airliner was shot down, Petrov's computer screen suddenly lights up. An alarm signal from the Russian satellite surveillance system begins to blare. On his display, he can see the signature signals from five incoming intercontinental ballistic missiles (ICBMs), launched from U.S. territory and streaking straight toward the heart of the Soviet mainland. His computer system confirms an imminent enemy missile strike. The computer system now instructs Petrov to send out a formal warning to headquarters. The computerized algorithm requires a "Launch" command be sent out to the entire air defense network along with the message. Petrov does not have the luxury of time to reflect on the situation, because the defense system has already calculated that on their current trajectory the ICBMs will reach Moscow in under twenty minutes.

Years later, Petrov recalled his predicament: "There was no rule about how long we were allowed to think before we reported a strike. . . . But we knew that every second of procrastination took away valuable time. . . . All I had to do was to reach for the phone; to raise the direct line to our top commanders—but I couldn't move."

He considered the data on his display. Throughout his military career, Petrov had always been instructed that if American strategic forces were to attack the Soviet Union, it would be with a withering, overpowering nuclear knockout punch aimed at eliminating the ability of Soviet forces to mount any effective retaliation. If this were the opening salvo in a global thermonuclear confrontation, then it seemed an odd one. There was a second nagging issue: the satellite detection equipment Petrov was using had only just been installed. It had been in use for less than a month. Not long enough to be thoroughly tested in rigorous practice drills. What little faith he might have had in the system was further shaken when it failed to return a satisfactory verification after he asked the computer to confirm the incoming radar signals. Petrov decided it was a false alarm. And he did not report any launch detection to Soviet headquarters.

Had Petrov followed his military training—or had the decision been left, as it might be today, to an AI algorithm—the launch detection would have been immediately passed up the chain of command. Down to the last nuclear expert, all agree that if the Soviet Union high command had received such information, then it would have ordered a full-scale, retaliatory counterattack on the United States. Without question. That would have prompted a full-scale, retaliatory "second strike" from the United States. A global thermonuclear war would have ensued, and so-called mutually assured destruction would have been the expected result.

It was only days later that experts confirmed that Petrov's equipment had, in fact, malfunctioned. Sunlight bouncing off high clouds over North Dakota had flashed just as a Soviet satellite was orbiting over-head, and it misread the signals as the traces left by a missile launch. In the West, the incident remained entirely unknown for another fifteen years. To this day, few people know how truly close the world came to annihilation in those early morning hours.

The question is, why? Why did Petrov take it upon himself to make this tough call? He could have just as easily said, "This is way above my pay grade. I'm going to kick this problem up to the high command." But he did not. Instead, he inserted his humanity and his reasoned approach into the process. His common sense plugged the hole in the swiss cheese.

After the incident, Petrov's actions were quickly reported to General Yuri Votintsev, the head of the Soviet Air Defense System. The records indicate the incident was "duly noted." Petrov received neither a repri-mand for failing to follow his standing orders nor a medal for saving the world. The malfunction of the missile defense system was a huge embarrassment to the Soviet military leadership. As a result, the episode was hushed up. Eventually, headquarters demanded an internal inves-tigation be carried out: it concluded that Petrov was guilty of "having insufficiently documented his findings at the time they happened." For this, he was reassigned to a position in a research lab that was deemed "of a less essential and sensitive nature." A year later, he was forced to retire from military service altogether. In the aftermath, he suffered a nervous breakdown.

No one would have ever heard of Petrov's heroism had it not been for General Votintsev's publishing his memoirs in 1998. It was the first time

U.S. intelligence or NATO analysts had even heard about it. Thirty-one years after the incident, a documentary film came out, titled *The Man Who Saved the World*. For the first time, the movie documented Petrov's courageous decision. Fortunately, Petrov lived long enough to see himself briefly celebrated for his decision to, as he put it, "do nothing."

Petrov's common sense stands out in stark contrast to the collective sleuthing, expertise, brinkmanship, and gadgetry of the military-industrial complexes of these two superpowers. I hope there were some notable moments in Petrov's life when he could pause to feel deep satisfaction and grace for what he did. Like when he heard children laughing at play. Or looked at young lovers walking hand in hand. Or he heard the clinking of glasses at a celebration. I hope that in those moments he could sweeten his remaining days by remembering he had made this all possible. He deserved that much. There is a small plaque now to commemorate him in the park that was built at what was ground zero in Hiroshima. The citizens of that city truly knew the collective horror Petrov had spared the world. It turns out we could use him now more than ever.

Elsa Kania, a fellow with the Technology and National Security Program at the Center for a New American Security, pointed out that invoking a kind of "Petrov's Principle" in the development of our advanced automated weapon systems may be more critical now than ever. The idea behind the principle is simple: Only a human being can decide to use a weapons system to launch an attack. Kania was attending a meeting in 2017 on autonomous killer drones. "Petrov's decision," she said, "should serve as a potent reminder of the risks of reliance on complex systems in which errors and malfunctions are not only probable, but probably inevitable."

We have entered a perilous period in human history where *AI could now make all of the necessary discretionary decisions on its own to launch a nuclear attack*. It could decide that it had gathered sufficient information to warrant such a response; it could then select which targets to choose and which weapons to launch. All of this, from start to finish, could be carried out without any human consultation. Petrov may yet become the patron saint of those who pray for protection from autonomous killer weaponry.

Terminal Trans-generational Events

An existential threat to humanity is any incident or process that has the potential to end in a so-called terminal trans-generational event. The words sound like a slogan on a spray can of roach killer. But the phrase is about us—you and me. And about our children and grandchildren. It means that we and all of our descendants are dead. It means our bloodlines have run dry.

In my clan, the Hamiltons, genealogical records have been preserved and handed down from one generation to the next. Why? Because each generation expected its lineage to be ensured. The Hamiltons can trace their bloodlines back forty generations to Halfdan Haraldsson, who lived from approximately 590 to 650 AD. Our ancestors fought alongside William the Conqueror and later, as Scots, next to Robert the Bruce. My ancestor David Hamilton fought against Cromwell's English forces at the Battle of Dunbar in 1650. He was captured and transported as a prisoner of war to the colonies in North America, where he was an indentured servant for a full decade before being able to secure his freedom. Fourteen generations later, I wear the same tartan he wore. But all of that tradition and cumulative history would be utterly destroyed. That's the real significance of a so-called terminal trans-generational event. It means history stops.

It is likely our species will face its "existential bottleneck" within this century. The terminal trans-generational event could take the form of a global thermonuclear war. Or global warming. Or a pandemic. Or the orchestrated elimination of our species by superintelligent technology. Or it could just be a meteor that collides with our planet and brings life as we know it to an end. A Global Catastrophic Risk Conference at Oxford University put the overall probability of human extinction before 2100 at 19 percent. A 2016 annual report of the Stockholm-based Global Challenges Foundation estimated the chances of an average American dying in a car accident are five times lower than his risk of perishing in a "human extinction event." Don't think that what we are discussing is theoretical, alarmist, or irrelevant. We are talking about the real risks your offspring and mine will face.

The brain is, first and foremost, an experiential mapping tool. It draws on the memories it has accumulated over the course of our lifetime—and, by proxy, those stored in all our libraries, hard drives, and servers—and then creates a map of choices. It assesses the risks, benefits, and alternatives available to us. But most important, the brain is seeking to plot a path with a high likelihood of success. The brain is also risk averse and more likely to falter when it encounters a novel situation of which it has little experience and for which it is ill-prepared. Existential threats, by their nature, are scenarios that leave little room for trial and error.

Of the potential extinction events we mentioned, all of them are of our own making. They're all man-made except for the meteor colliding with the planet, which truly is not our fault. It is the only natural disaster among them. We have had substantive growth in our fund of knowledge about the composition and behavior of comets. We even have an excellent idea of how a thermonuclear device could be used in space to cause a meteor to melt, veer off course, and avoid a collision with Earth.

When it comes to atomic bombs, however, the news is not as good. In 1946, there were 3 nuclear devices in the world. Now there are 13,082—give or take a few dozen that cannot be accounted for. Two extant treaties limiting atomic weapons have been allowed to lapse. The "Doomsday Clock" is a representation of current trends and forces in the world that make thermonuclear war more or less likely. The closer the clock is to the stroke of midnight, the worse the situation is looking. The clock is set each year by a panel of more than sixty scientists and experts across a host of disciplines. The clock started arbitrarily in 1947 at seven minutes to midnight. The best we have ever been able to push the clock back has been seventeen minutes to midnight; that was in 1991. Currently, the Doomsday Clock stands at ninety seconds to midnight—the closest it has ever been to the stroke of midnight in its seventy-five years. I don't know about you, but that makes me feel very uncomfortable.

Similarly, we can look at global warming and say that clock is ticking too. All the studies and scientific investigations plainly map out where we are headed: the planet is getting too hot. And it is clear the planet is headed for catastrophe. We should be planning accordingly. In that regard, we may be suffering from the environmental equivalent of "boiling frog syndrome": persistent denial in the face of a bubbling, overheating conclusion.

Our problem is that we have this fallacy of Earth as a fundamentally inexhaustible resource. Nothing could be further from the truth. Earth Overshoot Day is the designated day in the calendar year when human consumption outstrips the planet's production of natural resources. In 1970, Overshoot Day did not exist. By 1997, it had reached October 1. In 2019, it was July 29. That is the 210th day of the year. It means that for the remaining 155 days of the year, we are drawing down on Earth's natural resources in an unsustainable way.

Regarding the dangers of a pandemic, we only need to look around us at the global devastation left behind by the COVID-19 crisis. I can tell you that despite all the advances we have made in medical science in the last hundred years, we made almost the identical set of mistakes with the COVID-19 outbreak that we did with the influenza pandemic of 1918. This means that despite all the medical advances in a century, human errors were the same. The denial, the finger-pointing, the procrastination, the unwillingness to understand the global context, the delay in central control and preventive measures: all were unchanged despite one hundred years of history in between. Karma is a bitch.

As far as AI is concerned, despite the lip service that many countries make, there has not even been a *draft* of an agreement in front of the United Nations to get the nations to help establish guidelines to safeguard the development of AI. In his book *Army of None,* the military analyst Paul Scharre points out that currently more than thirty countries have autonomous defensive weapons systems that are operating at speeds so fast that humans cannot respond. Mary Wareham, the director of the Arms Division of Human Rights Watch, believes AI-driven autonomous weapons will be upon us not in a few decades but in a few years—at the most.

In 2020, Mohsen Fakhrizadeh, Iran's top nuclear scientist and a commander in Iran's Islamic Revolutionary Guard Corps, was being escorted outside Tehran by a security convoy. A nondescript pickup trip lay abandoned along the convoy's route. It was off at some distance from the road and therefore not considered a risk by his security detail. However, a very accurate machine gun was mounted in the truck and camouflaged to avoid detection. The device used AI for facial recognition and was operated via satellite. As a result, it was able to assassinate Fakhrizadeh in his vehicle without touching his wife, who sat right next to him.

The situation surrounding AI-controlled weaponry is indeed dire. The Russian president, Vladimir Putin, stated in 2017, "Artificial intelligence is the future, not only for Russia, but for all humankind. It comes with colossal opportunities, but also threats that are difficult to predict today. Whoever becomes the leader in this sphere, will become the ruler of the world." Sometimes, it is hard to know if that is an admonition or an exhortation.

There have already been numerous terrorist attacks using autonomous, AI-enhanced weaponry. It is tailor-made for asymmetric warfare. In September 2019, Houthi rebels in Yemen used GPS-programmed targeting to launch an attack on two large Saudi Arabian petroleum facilities at Abqaiq and Khurais. The rebels used twenty-five store-bought drones to attack a large battery of MIM-104 Patriot missiles located next to the facility at Abqaiq. It was completely inoperable against the attack because the drones were flying so low and so slow as to avoid detection. The installation of the missile battery cost hundreds of millions of dollars, and each Patriot missile in it cost more than $3 million. The drones were purchased out of a mail-order catalog and cost less than $200 a piece. Asymmetric warfare, indeed. The attack on the oil fields succeeded at cutting global oil production by 5 percent. ISIS has launched three similar attacks in Syria. Saint Petrov, we beseech you to hear our prayers.

But not all the news is bad.

Has Humanity Improved?

Yes. Progress in human affairs is real. Take, for example, the massive mobilization of assets and materials that occurred under the Marshall Plan at the end of World War II. It was an unprecedented effort to infuse money and materials to quickly rebuild the infrastructure and economies of our vanquished enemies, Germany and Japan. It turned both nations into our stalwart economic and strategic partners for seventy-five years after the war concluded.

In a recent TED talk, Steven Pinker pointed out that yearning for "the good ole days" is really a testament to false memories about what those days were really like. Homicide rates in the United States today are 5.3

per 100,000 individuals compared with 8.5 per 100,000 citizens thirty years earlier. Poverty rates in America stand (or at least *stood* before the COVID-19 pandemic) at 7 percent as compared with 12 percent thirty years ago. Pollution production has fallen by fourteen million tons of particulate matter and sixteen tons of sulfur dioxide over the last three decades. Nor are such improvements limited to America. Poverty rates in the world stand at 10 percent as compared with 37 percent thirty years earlier. In 1960, the global literacy rate was less than 40 percent. It had risen to 85 percent by 2015. The average weekly work hours have decreased by 50 percent in the last century. Since 1990, mortality secondary to pollution has been halved. In the Roman Empire, slavery rates were as high as 30 percent, while currently global slavery indices are less than 1 percent.

Can Humanity Save Itself?

Nicholas Christakis is trained both as a physician and as a sociologist and is the director of the Human Nature Lab at Yale. He is also a favorite writer of mine. In his book *Blueprint: The Evolutionary Origins of a Good Society*, Christakis looks at what he calls "natural experiments." "Natural experiments," he explains, "allow scientists to circumvent practical impediments, mitigate ethical obstacles (like killing spouses), and study large-scale phenomena that are impossible to replicate (such as the effects of military invasions)." Christakis makes a brilliant choice of using shipwrecks as the natural experiment to evaluate how humanity rallies its resources (or fails to) when stranded and removed from society. He explains,

> Survivor camps established after shipwrecks provide fascinating data about the societies that groups of people make when it's left up to them, about how and why social order might vary, and about what arrangements are the most conducive to peace and survival. An archipelago of shipwrecks, formed over centuries, more or less at random, has resulted in people participating, unintentionally, in multiple trials of this experiment.

Christakis points to the wreck of the *Medusa* in 1816 as a first example. The French frigate was sailing to reestablish a colony in Senegal as soon as the Bourbon monarchy was restored to the throne after the exile of Napoleon. The ship had 400 souls on board. Command of the vessel had been bestowed upon a naval officer as his reward for remaining loyal to the French crown throughout the reign of Napoleon and then returning to serve his king. He had scarcely set foot on a ship in twenty years. The man was incompetent. He mistakenly navigated the ship more than 160 miles off course and wrecked it off the coast of Mauritania. About 250 passengers made it to the rowboats. About 150 individuals had to climb onto a hastily constructed raft that was to be towed behind the boats. However, the raft was bulky, badly overloaded, and difficult to tow.

After just a few miles, the captain made the decision to cut the raft loose, leaving the men on it with no means of navigation in the open sea. On the raft, they had one barrel of water (lost in a scuffle), one bag of biscuits (lasted one day), and six barrels of wine. Of the men on the raft, scores were washed overboard by high waves. Some became drunk and began to mutiny and were shot dead by some of the ship officers on board. As supplies quickly dwindled, any injured individuals on the raft were tossed into the sea. And when the food was gone, the strongest among them turned to cannibalism. A ship came upon the raft thirteen days later; only fifteen men survived.

Compare the *Medusa*'s experience with that of the *Doddington,* a cargo square-rigger sailing from Dover, England, to Bombay, India, in 1755. After rounding the Cape of Good Hope, the ship crashed on July 17, 1755, into a rock in Algoa Bay at night and quickly sank. Of the 270 souls on board, only 23 survived. The castaways made their way to an island. Several of the crew helped those who were hurt swim to shore. Once ashore, they did their best to attend to the wounded. The island was a prime seabird nesting site and rookery. The castaways subsisted on a diet of birds, eggs, and fish for seven months. The ship's carpenter was among the survivors, and he oversaw the construction of a sloop made from timber they had salvaged from their wrecked ship. They christened the boat *Happy Deliverance.* On February 16, 1756, the remaining 22 survivors set sail from their island and traveled up the coast of Africa for two months until they landed safely in present-day Mozambique.

To put the differences in outcome between these two groups of ship-wreck survivors into perspective, let's review their respective mortal-ity statistics. For the raft of the *Medusa,* there were 15 survivors of the roughly 150 people on the raft. That would represent a 90 percent mortal-ity rate over the course of thirteen days! Compare that to the Dodding-ton: where 22 souls survived out of 23, representing a mortality rate of only 4 percent over a much more extensive period of time, namely, nine months! How do we explain such vastly different survival rates between the two wrecks? Christakis points to what he identifies as "the social suite"—namely, the intellectual and emotional capacities that lie at the heart of all human societies. The "suite" can be summarized into five substantive areas:

CAPACITIES OF THE SOCIAL SUITE

1. The capacity to recognize individuals' identity
2. The expression of love for partners and offspring
3. The establishment of friendships and social networks
4. Cooperative endeavors within mildly hierarchical organization
5. In-group bias

The features of this social suite should be readily recognizable as reflections of specific properties of brain function that manifest them-selves in our private and social lives. They arise from the all-too-familiar evolutionary tension generated by our central nervous system to find our own intimate in-group. We can say, short of severe psychopathology, all human beings seek to establish their in-group, whether it be a traditional family unit, a team, a military squad, a church group, or even a prison gang. We are driven to establish a tribe to which we can belong and where we will thrive. In whatever format we establish our sense of family, the brain's capacities for love, empathy, and compassion are powerful and substantive.

What Christakis found was that shipwreck survivors were quickly guided into either a communal or a predatory spirit. Often the key (as in the wreck of the *Doddington*) lay in some early act of altruism and/ or charity, often in the very first minutes or hours after disaster struck. In the case of the *Medusa,* the tone of the shipwreck (and its subsequent "rescue") were established by the decision to cast 150 people adrift on a raft in the seas off the coast of Africa. In many of the shipwrecks, kind and respectful officers often helped establish a civil, courteous tone. Often groups would even abandon the naval chain of command to institute election of leaders by democratic governance. Initial acts of altruism got the shipwrecked started on the right path. By contrast, an early act of cruelty sent the survivors into a downward spiral of exploitation and ruthlessness.

These two tendencies vie with each other in all of us. But it matters how we start. The tone we set as we initiate our endeavors can determine their future success or failure. Look at the radical differences in the outcomes of these two shipwrecks. One road leads to a survival rate of 10 percent while the other reaches a 96 percent survival rate! So, it matters how your group starts! The first steps may be the most important we take. It matters what tone the group sets and what examples the leaders provide. It matters what values the group installs and embraces. And what principles do they live by? We must remind ourselves of these notions because, like the shipwreck survivors, the path our country and society choose—whether we govern by the principles of inclusion or exclusion, or are guided by securing the common good over pursuit of ruthless individual advantage—may all impact how many we rescue.

Better Angels

When Abraham Lincoln took office in 1861, he grasped the fundamental conflict between including and excluding segments of society. He knew that seven states had already withdrawn from the Union over the issue of slavery, and he knew more would follow. Yet in his inaugural address, he urged listeners to remember, "We are not enemies, but friends. We must not be enemies. Though passion may have strained it must not break our bonds of affection. The mystic chords of memory . . . will yet swell . . . when again touched, as surely they will be, by the better angels of our nature." It is those angels that bind the social suite together and bestow upon us the benefits of its blessings.

Our current understanding of the neuroscience of goodness is that it depends on our ability to identify commonality. How do we recognize commonality? It moves us. I carry in my wallet two photographs. They are like symbolic smelling salts to snap me out of it when I feel sorry for myself. One is the Pulitzer Prize–winning photograph taken by Kevin Carter of a young girl trying to make her way to a food relief station during the famine in Sudan while a vulture follows close behind, betting that the child will not make it. The second is a picture of a three-year-old boy whose body washed up on a beach in Turkey. He is a dead refugee whose family tried to escape the violence in Syria. He perished at sea as he and his family sought asylum. These two photos remind me of the dangers of inadequate compassion: rejection, omission, and exclusion. And death.

Humanity struggles at each stage of civilization to reimagine the scale of its in-groups. But the progress of human society can be traced to its ability to consistently recalibrate its sense of inclusion. The pandemic, global warming, and even world trade have all demonstrated that the entire planet—and all of its life-forms—must be included in our calculations. We have recently seen the stirrings of a great social shift in how we delineate our in-groups. The prominence and global propagation of movements like Black Lives Matter, the #MeToo movement, and the assertion of the rights of the LGBTQ community are all manifestations of the same aspiration to grow by becoming more inclusive and to expand the embrace of our in-group. If history teaches us one thing, it is that "the better angels of our nature" are rarely wrong, and we falter only when we fail to be moved by them.

The Fermi Paradox

Enrico Fermi was one of the great atomic physicists of the twentieth century. He took note of an apparent contradiction, known as the Fermi paradox. Fermi demonstrated that the mathematical odds of finding other forms of intelligent life elsewhere in the universe were quite high. And yet, Fermi pointed out, no sign of extraterrestrial intelligent life has ever been found. He came to the conclusion the paradox could be explained in one of two ways. The first, Fermi reasoned, was the less likely of the two hypotheses: namely, that the other forms of intelligent life in the universe didn't want to be discovered and were hiding effectively from our view. But his second hypothesis was that intelligent life was not to be found simply because sooner or later it ends up wiping itself out. Life becomes the victim of an abundance of intelligence married to a paucity of wisdom.

The reciprocity between our emotions and our relationships is similar to the entanglement between the physical world and the perceptual one. What we look for in people is what we will discover. We saw this same mutuality in oxytocin release and the mirror neuron system. As Daryl Davis proved, we can find friends among our enemies when we no longer see them as enemies.

In the introduction, I asked us to consider the possibility that one of the "objectives" of the universe might be to foster the evolution of intelligent life so it can, in turn, recognize and embrace the creative purpose at the heart of the universe. Could the purpose of creation be to recognize creation?

If so, we need more time—something that may be in short supply. If we are to prove the Fermi paradox wrong, then we have to be bold enough to envision an inclusive planet. If we are to achieve the purpose that might have been unfolding patiently over thirteen and a half billion years across the backdrop of the stars, we need to recapture our sense of awe.

Awe is what we often experience in the face of nature. I have experienced it myself many times: on the ocean, in a hushed forest, or from the summit of a high peak. Nature casts its spell by allowing us to be stirred by its scope and scale. Awe is, I believe, the best reset button we may have as a species to reaffirm we live *in* the world and not just *on* it.

Awe has repeatedly inspired our astronauts. Few of us have had the opportunity to see such a perspective—at least for now. But half a century ago, the astronaut William Anders was looking out the window of the Apollo 8 spacecraft while it was in lunar orbit. His job was to photograph potential sites where later Apollo missions could land on the moon. At one point, he looked up and saw Earth rising in the depths of space over the lunar horizon. He snapped a photograph—one that has been called the defining snapshot of the twentieth century (see figure 15).

Frank Borman, the commander of the Apollo 8 mission, later commented on the sense of wonderment and scale that he personally experienced looking back at our home planet: "The view of the Earth from the Moon fascinated me—a small disk, 240,000 miles away. . . . Raging nationalistic interests, famines, wars, pestilence, don't show from that distance." He added, "The isolation of Earth . . . and the closeness of humanity . . . I wish more people could focus on that."

From space, the scale of life on our home planet looks small and fragile. The French philosopher Blaise Pascal wrote, "The eternal silence of these infinite spaces fills me with dread." It should because, from this vantage point, the isolation of our planet demands we embrace the finality of the common destiny it imposes upon our human family.

Acknowledgments

A book is always a team effort. No matter how many people you try to thank for their contributions to that effort, you will always fall short, but as a writer you always try to remember and thank everyone.

First and foremost, I have to thank my wife, Jane. She met me when I was working as a janitor, and when I told her that I wanted to become a brain surgeon, she was the first person who didn't laugh at me. She never laughed. Instead, she just said, "Why not? What's stopping you?" And I am eternally grateful for her confidence in me and the endless font of encouragement and inspiration she has been throughout my life. For more than half a century, she has always known I loved her, but she also knew I was in love with the brain, which can be a cruel mistress sometimes. My family made many sacrifices because of my passion for the brain. So often, my children's Little League game or school play would be eclipsed by my having to go to the hospital to perform emergency surgery. They always forgave me, as hard as that could sometimes be. My mom passed away during the writing of this book, but she helped me to write it. When I was very young, she always fed me books about scientists and anatomists. We were not a wealthy family by any measure, but my mom made sure we always had books about science, and these were filled with heroes I wanted to emulate one day.

I have had so many mentors in my career that it is hard to acknowledge them all. As an English major at Ithaca College, I was fortunate to spend two full years studying creative writing with Rod Serling, the creator of *The Twilight Zone* and one of the greatest creative forces I have ever met. He insisted that for every semester I studied with him, I take the same number of credits studying Shakespeare with Professor John Harcourt. The juxtaposition of studying Shakespeare's plays while reading the science fiction of the age was heady stuff. Through Mr. Serling, our class was introduced to Ray Bradbury, Isaac Asimov, Kurt Vonnegut, and Harlan Ellison, among the most inspiring writers. Mr. Serling was the epitome of a great storyteller, and I am forever in his debt for giving me the confidence to believe in my own storytelling abilities.

I have been blessed with an extraordinary cadre of some of the best mentors and teachers of the twentieth century medicine for inspiration and training: Alvin Poussaint, Francis Moore, Judah Folkman, Dan Federman, Lachlan Forrow, Nicholas Zervas, Paul Chapman, Roberto Heros, Robert Ojemann, Charles Poletti, Bill Sweet, Ray Kjellberg, John Shillito, Larry Borges, Brooke Swearingen, Bob Martuza, Rees Cosgrove, H. T. Ballantine Jr., Ned Cassem, Ray Adams, C. Miller Fisher, E.P. Richardson, Tessa Hedley-White, L. Philip Carter, Robert Spetzler, Joe Zabramski, Andy Shetter, Ron Weinstein, Jim Dalen, Joe Alpert, Fayez Ghishan, Rich Carmona, and my neurosurgical faculty partner of thirty years, Marty Weinand, and too many others than I can list here. I am grateful for all the wonderful relationships I have had with my fellow residnts and colleagues at the Massachusetts General Hospital and, later, at the University Medical Center at the University of Arizona Health Sciences Center in Tucson. Every single one of them, I realize now, reached out to me to include me in their-in group, to invite me to join hands with their circle of friends, family, and colleagues. I have known of no other way to thank them than to pay it forward.

Regarding writing *Cerebral Entanglements,* I was fortunate to have a close friendship with my agent, Laura Yorke, who is a wonderfully gifted writer and editor in her own right. Laura spent endless hours encouraging me and helping me rewrite and reshape the book through dozens of revisions. Rewriting can be complicated, but I never felt alone in my foxhole for one second, having to handle the onslaught of yet another rewrite, because Laura was always at my side. There are simply no words to express what it means to have someone who believes fiercely in your work; it is like sharing it with another soul.

I am indebted to Vicky Wilson, Marc Jaffee, Belinda Yong, Lisa Lucas, and Lisa Kwan for their editorial assistance. I owe an enormous debt of gratitude to Anthony Ziccardi at Post Hill Press (PHP), who immediately threw the full force of his support behind this book and understood its potential impact. I also want to thank Mike Hagler and Caitlin Burdette of PHP for tirelessly tweaking the copy to get this book to press. I want to thank Sam Mitchell, Sani Mendelson, and Sarah Payne of Hilsinger-Mendelson, Inc. with their kind and compassionate focus one the book's publicity.

A big thank you to Kate Gardiner from the University of Arizona Biomedical Photography Department for her beautiful photographs, and Ricky Bergeron, my buddy through more than a decade and a half of work with the Division of Biomedical Communications at the University of Arizona, for taking time out from his vacation to help me with some of the images in the book when I was backed up against a deadline.

I would be remiss if I did not mention my secret weapons in all this: first, is Kris Spinning, my website designer and social media guru who has helped me with almost every book I have written. Secondly is my office manager, Kate Lamkin. *Cerebral Entanglements* is the third book that Kate has helped me with, and she understood that this book represented a massive undertaking but also one that was very dear to my heart. This is a summation of five decades of learning about "the most complicated object in the whole universe." I am eternally grateful for her patience and kindness with the endless multitasking and tweaking that I required to see this project to its conclusion.

The COVID-19 pandemic wreaked havoc with our lives and the progress of the manuscript. My mom passed away in the first months of the pandemic. She was proud of my being a writer, and she always had a copy of each book I had written on a side table next to her favorite armchair. I wish she had gotten to see this one. A year later, Vicky reached a well-deserved retirement and I felt like I had been thrown out of a plane, clutching my manuscript with no parachute. In free fall, waiting to slam into the earth.

Someone did finally grab me, and then held onto me and my manuscript, and pulled the ripcord so our chute deployed: Lisa Lucas. She read my book. She got it. She gave me some suggestions to make it better. I felt completely at ease asking questions about passages in the book that were important to me or I thought might be controversial. Lisa is a publishing thoroughbred. She started as an intern with *Vibe* magazine at the age of fifteen. She worked her way diligently through the ranks of the publishing world and in 2016 was appointed to the prestigious position of executive director of the National Book Foundation.

Lisa came to look at the book with really new eyes. Sometimes, as a surgeon, you can find yourself "stuck" in the operation. It seems as if you become trapped in the dissection, and each move you want to make only appears to make matters worse. That's when you call a partner or a colleague and ask if they can come down to the operating room, scrub into the case, and "assist" you. And every time, that extra set of eyes looking at the problem and the extra pair of hands helping, is all it takes to break the logjam and the case suddenly is moving froward again. That new insight from a new pair of eyes, a different set of hands working the case, that is what Lisa brought to this book. There was nothing that she was not willing to reconsider or reimagine about the manuscript. She also stacked the deck in my favor by putting Ben Shields in her office in charge of helping me mould (hammer might be a better word) the manuscript into shape. I have spent more than half a century in love with the brain and this was the most important book to me I have ever written. Putting up with a passionate, egocentric brain surgeon who is writing a manuscript with you is like being told your new roommate is a psychotic, serial killer. You're pretty sure that running away could save your life and sticking it out might imperil it. Thank you, Ben, for sticking it out. I also need to thank my copy editor. Ingrid Sterner. You form a weird relationship with your copy editor because you only know each other by the comments you leave each other. It's like dating someone but the only contact you have are the Post-it notes you leave for each other on the fridge. But there are so many of them that you do get to know each other and appreciate how much that copy editor cares about helping you get the detail right.

A big thank-you to Ricky Bergeron, my buddy through more than a decade and a half of work with the Division of Biomedical Communications at the University of Arizona, for taking time out from his vacation to help me with some of the images in the book when I was backed up against a deadline. And I need to also thank Debbie Goodsite who labored with me for days ensuring that the myriad routes by which we obtained permission for our figures still gave us the best images humanly possible. I feel indebted to her since the images of the brain can often be the heart of the story. Also, a big hug and thank you to Kris Hanning for taking photos too.

I would be remiss if I did not mention my secret weapon in all this: my office manager, Kate Lamkin. *Cerebral Entanglements* is the third book that Kate has helped me with and, by far, the biggest. Kate, however, understood that this book represented a massive undertaking but also one that was very dear to my heart. This is a summation of five decades of learning about "the most complicated object in the whole universe." I am eternally grateful for her patience and kindness with the endless multitasking and tweaking that I required to see this project to its conclusion.

Finally, during the course of career as a brain surgeon, there have been so many patients with whom I shared part of their lives and they part of my own. Our lives always remain intermingled thereafter. Their stories and courage remain with me and always inspire me There is no greater faith a person can express in another than to say: "I am entrusting you to operate on my brain, the seat, the guardian, the temple of my being. Promise me you will preserve my functions, my experiences, my memories, my life to the utmost of your abilities." For that trust, I am forever grateful.

Notes

Introduction

3   "There is no scientific study": Francis H. Crick, "Thinking About the Brain," *Scientific American* 241, no. 3 (1979): 232.

10 On May 14, 1964: Kevin Crosby, "Jose Delgado and Mind Control," video, Feb. 8, 2013, YouTube, www.youtube.com/watch?v=eK2Hopm5s_c.

10 Dr. Delgado stepped out: John A. Osmundsen, "'Matador' with a Radio Stops Wired Bull; Modified Behavior in Animals Subject of Brain Study," *New York Times,* May 17, 1965.

13 utopian society of the future: José M. R. Delgado, *Physical Control of the Mind: Toward a Psychocivilized Society* (New York: Harper & Row, 1969).

1. On the Nature of Consciousness

17 "I know it": Peter Lattman, "The Origins of Justice Stewart's 'I Know It When I See It,'" *Wall Street Journal,* Sept. 27, 2007.

17 "'you,' your joys": Rosie Mestel, "Renowned DNA Scientist Saw Life as It Is," *Los Angeles Times,* July 30, 2004.

20 With the fMRI: Karina Quevedo et al., "The Neurobiology of Self-Face Recognition in Depressed Adolescents with Low or High Suicidality," *Journal of Abnormal Psychology* 125, no. 8 (2016): 1185–200, doi.org/10.1037/abn0000200.

20 "When . . . [depression] comes": Andrew Solomon, *The Noonday Demon: An Atlas of Depression* (New York: Scribner, 2014), 15.

25 "a merely physical life": Steven Laureys et al., "Brain Function in the Vegetative State," *Acta Neurologica Belgica* 102, no. 4 (Dec. 2002): 177.

27 "Not being able": Rebecca Morelle, "I Felt Trapped Inside My Body," BBC News, Sept. 7, 2006.

35 "Switching from task to task": Victor Imbimbo, "Is Multitasking Actually a Myth?," *HuffPost,* Jan. 24, 2014.

35 A public firm: Travis Bradberry, "Multitasking Damages Your Brain and Career, New Studies Suggest," *Forbes,* Dec. 10, 2021.

38 "Consider again that dot": Carl Sagan, address at Cornell University, Oct. 13, 1994, archived at www.bigskyastroclub.org/pale_blue_dot.html.

2. On the Nature of Affection and Trust

44   substantial amounts of oxytocin: Association for Psychological Science, "Level of Oxytocin in Pregnant Women Predicts Mother-Child Bond," news release, Oct. 15, 2007.

44   "primes" complex maternal behaviors: Lydia Denworth, "The Good and Bad of Empathy," *Scientific American,* Dec. 1, 2017.

44   rat mothers who: Cort A. Pedersen and Arthur J. Prange Jr., "Induction of Maternal Behavior in Virgin Rats After Intracerebroventricular Administration of Oxytocin," *Proceedings of the National Academy of Sciences* 76, no. 12 (1979): 6661–65, doi.org/10.1073/pnas.76.12.6661.

44   evolutionary selection: Mary D. Salter Ainsworth, "Object Relation, Dependency, and Attachment: A Theoretical Review of the Infant-Mother Relationship," *Child Development* 40 (1969): 969–1025.

45   Romance produces: Gary Stix, "Fact or Fiction? Oxytocin Is the 'Love Hormone,'" *Scientific American,* Sept. 8, 2014.

45   Furthermore, the higher: Association for Psychological Science, "Level of Oxytocin."

45   This lowers circulating levels: Marek Jankowski, Tom L. Broderick, and Jolanta Gutkowska, "The Role of Oxytocin in Cardiovascular Protection," *Frontiers in Psychology,* Aug. 25, 2020, doi.org/10.3389/fpsyg.2020.02139.

46   They also exhibited: Ross Andersen, "The Case for Using Drugs to Enhance Our Relationships (and Our Break-Ups)," *Atlantic,* March 9, 2013.

46   Couples who experience: John Tully et al., "The Effect of Intranasal Oxytocin on Neural Response to Facial Emotions in Healthy Adults as Measured by Functional MRI: A Systematic Review," *Psychiatry Research: Neuroimaging* 272 (2018): 17–29, doi.org/10.1016/j.pscychresns.2017.11.017.

47   "You know this": Dwane Brown, "How One Man Convinced 200 Ku Klux Klan Members to Give Up Their Robes," NPR, Aug. 20, 2017.

50   Dr. Charles Nelson: Jon Hamilton, "Orphans' Lonely Beginnings Reveal How Parents Shape a Child's Brain," NPR, Feb. 24, 2014.

51   Their brains could no longer: Ibid.

52   The emperor ran: "Emperor Frankenstein: The Truth Behind Frederick II of Sicily's Sadistic Science Experiments," History Answers, Sept. 6, 2017.

52   The RAND Corporation: Lynn A. Karoly et al., "Early Childhood Interventions: Benefits, Costs, and Savings," RAND Corporation Research Brief no. RB-514 (1998), doi.org/10.7249/RB5014.

53   Most beneficiaries were: Ben L. Kail and Marc Dixon, "The Uneven Patterning of Welfare Benefits at the Twilight of AFDC: Assessing the Influence of Institutions, Race, and Citizen Preferences," *Sociological Quarterly* 52, no. 3 (2011): 376–99, doi.org/10.1111/j.1533-8525.2011.01211.x.

54   "In 21 states": Shawn Fremstad, "Budget Cuts Are Putting More Kids in Foster Care," *Talk Poverty,* Feb. 23, 2018.

54   The traumas involved: Ryu Takizawa et al., "Bullying Victimization in Childhood Predicts Inflammation and Obesity at Mid-life: A Five-Decade Birth Cohort Study," *Psychological Medicine* 45 (2015): 2705–15.

54   Evaluating a large group: Dante Cicchetti, Elizabeth D. Handley, and Fred A. Rogosch, "Child Maltreatment, Inflammation, and Internalizing Symptoms: Investigating the Roles of C-Reactive Protein, Gene Variation, and Neuroendocrine Regulation," *Development and Psychopathology* 27 (2015): 553–66.

55   The latter are known: Andrea Danese and Stephanie J. Lewis, "Psychoneuroim-

munology of Early-Life Stress: The Hidden Wounds of Childhood Trauma?," *Neuropsychopharmacology* 42 (2017): 99–114, doi.org/10.1038/npp.2016.198.

55 significant mortality: Joel Gelernter, "The Epigenetics of Child Abuse," *Yale Medicine Magazine,* Nov. 12, 2014.

55 The significance of this finding: Karl M. Radtke et al., "Transgenerational Impact of Intimate Partner Violence on Methylation in the Promoter of the Glucocorticoid Receptor," *Translational Psychiatry* 1, no. 7 (2011): e21, doi.org/10.1038/tp.2011.21.

3. On the Nature of Romance

58 "Tonight I can write": Pablo Neruda, *The Poetry of Pablo Neruda* (New York: Farrar, Straus & Giroux, 2015), 19.

58 "some people would say": Sarah Pinto quoted in "How Do You Define Romantic Love?," *This,* [AU: Here and throughout the notes, please don't use "n.d." with online material. Instead, please give access date.]n.d., tinyurl.com/44849ame, accessed Feb. 5, 2019.

62 Love not only serves: Denworth, "Good and Bad of Empathy."

63 one could even call: Helen Fisher, *Why We Love: The Nature and Chemistry of Romantic Love* (New York: Henry Holt, 2004).

63 "Immature love says": Erich Fromm, *The Art of Loving* (New York: Harper & Brothers, 1956), 46.

64 "someone is camping": Helen Fisher, "The Brain in Love," TED talk, March 4, 2014.

64 "In the right proportions": Harvey Milkman, "What Makes Tigers Jump?," *Psychology Today,* Dec. 4, 2009; L. Slater and J. Cobb, "True Love: Scientists Say That the Brain Chemistry of Infatuation Is Akin to Mental Illness—Which Gives New Meaning to"Madly in Love," *National Geographic,* 9, no. 2 (2006): 32.

65 Neuroimaging demonstrates: Helen Fisher, "Romantic Love: Can It Last?," HelenFisher.com, accessed Jan. 29, 2022.

65 spouses in successful: Bianca P. Acevedo et al., "Neural Correlates of Long-Term Intense Romantic Love," *Social Cognitive and Affective Neuroscience* 7, no. 2 (Feb. 2012): 145–59, doi.org/10.1093/scan/nsq092.

65 as do newlyweds   ds: "Love Can Last: SBU Imaging Study Shows Brain Activity of Those in Love Long Term Similar to Those Newly in Love," *Stony Brook University News,* Jan. 7, 2011.

65 To answer this question: Helen Fisher, "The Science Behind Maintaining a Happy Long-Term Relationship," video, *Big Think,* Feb. 11, 2016.

66 We saw activity: J. Reid Meloy and Helen Fisher, "Some Thoughts on the Neurobiology of Stalking," *Journal of Forensic Sciences* 50, no. 6 (2005): 1 9, doi.org/10.1520/jfs2004508.

67 "Lloyd: What do you think": *Dumb and Dumber,* directed by Peter Farrelly, 1994.

68 High dopamine output: Meloy and Fisher, "Some Thoughts."

68 The case of E.F.: Laura Richards, "Rejected, Love-Obsessed, and Erotomanic: Inside the Disturbed Mind of a Stalker," *Telegraph,* May 8, 2017.

68 The sad news is: "Stalking Statistics," Stalking, Women and Gender Advocacy Center, Colorado State University, wgac.colostate.edu/support/stalking.

68 Ninety-eight percent of us: Dianne Grande, "Unrequited Love," *Psychology Today,* March 13, 2020.

69 Parts of the template: Jessica Herman, "Love at First Sight: Completely Crazy or Totally Possible?," *Cosmopolitan,* March 8, 2013.

69 If they are a very close match: Aaron Ben-Ze'ev, "Is Love at First Sight Possible?," *Psychology Today,* Nov. 17, 2013.

69   nearly half of all marriages: Ibid.

70   However, decades later: J. S.Hyde, R. S. Bigler, D. Joel C. C. Tate, and S. M. van Anders, "The Future of Sex and Gender in Psychology: Five Challenges to the Gender Binary," *American Psychologist* 74, no. (2019): 171, doi.org/10.1037/amp0000307.

70   Variation in biological: V. S. Helgeson, "Methods and History of Gender Research," in *Psychology of Gender* (New York: Routledge, 2020), 36–75, doi.org/10.4324/9781003016014.

71   brain imaging studies: Ivanka Savic and Per Lindström, "PET and MRI Show Differences in Cerebral Asymmetry and Functional Connectivity Between Homo- and Heterosexual Subjects," *Proceedings of the National Academy of Sciences* 105, no. 27 (2008): 9403–8, doi.org/10.1073/pnas.0801566105.

72   A research paper: Jerome A. Cerny and Erick Janssen, "Patterns of Sexual Arousal in Homosexual, Bisexual, and Heterosexual Men," *Archives of Sexual Behavior* 40, no. 4 (2011): 687–97, doi.org/10.1007/s10508-011-9746-0.

73   women in their twenties and thirties: Jean M. Twenge, Ryne A. Sherman, and Brooke E. Wells, "Changes in American Adults' Reported Same-Sex Sexual Experiences and Attitudes, 1973–2014," *Archives of Sexual Behavior* 45, no. 7 (2016): 1713–30, doi.org/10.1007/s10508-016-0769-4.

73   come out later: Institute of Medicine, Committee on Lesbian, Gay, Bisexual, and Transgender Health Issues and Research Gaps and Opportunities, *The Health of Lesbian, Gay, Bisexual, and Transgender People: Building a Foundation for Better Understanding* (Washington, D.C.: National Academies Press, 2011).

73   structural differences in the brains: Antonio Guillamon, Carme Junque, and Esther Gómez-Gil, "A Review of the Status of Brain Structure Research in Transsexualism," *Archives of Sexual Behavior* 45, no. 7 (2016): 1615–48, doi.org/10.1007/s10508-016-0768-5.

74   transgender clinics: Azeen Ghoryashi, "Texas Youth Gender Clinic Closed Last Year Under Political Pressure," *New York Times*, March 8, 2022.

74   this issue usually: Nienke M. Nota et al., "Brain Functional Connectivity Patterns in Children and Adolescents with Gender Dysphoria: Sex-Atypical or Not?," *Psychoneuroendocrinology* 86 (2017): 187–95.

74   experience issues: Malvina N. Skorska et al., "A Multi-modal MRI Analysis of Cortical Structure in Relation to Gender Dysphoria, Sexual Orientation, and Age in Adolescents," *Journal of Clinical Medicine* 10, no. 2 (2021): 345, doi.org/10.3390/jcm10020345.

74   not a single agency: Emanuella Grinberg and Dani Stewart, "Three Myths That Shape the Transgender Bathroom Debate," CNN, March 7, 2017.

75   interim results: George E. Vaillant, *Triumphs of Experience: The Men of the Harvard Grant Study* (Cambridge, Mass.: Harvard University Press, 2012); J. W. Shenk, "What Makes Us Happy?," *Atlantic*, July 10, 2020.

75   being closely connected: Robert Waldinger, "What Makes a Good Life? Lessons from the Longest Study on Happiness," TED talk, www.youtube.com/watch?v=8KkKuTCFvzI.

75   "Happiness is love": George E. Vaillant, "Happiness Is Love: Full Stop," MS (2012), www.duodecim.fi/xmedia/duo/pilli/duo99210x.pdf. See also Scott Stossel, "What Makes Us Happy, Revisited," *Atlantic*, Feb. 19, 2014.

4. On the Nature of Empathy, Altruism, and Kindness

77   "Three things in human life": Leon Edel, *Henry James: The Master, 1901–1916* (New York: Avon Books, 1978), 124.

78   "I am going to": George E. Vaillant, *Spiritual Evolution: A Scientific Defense of Faith* (New York: Broadway Books, 2008).

80   game for money: David Robson, "The Reason Why 'Everyday Heroes' Emerge in Atrocities," BBC, Nov. 17, 2015.

80   Carnegie Hero Fund: "Historical Timeline," April 20, 2023, Carnegie Hero Fund Commission, www.carnegiehero.org/about/history/historical-timeline/.

80   "If I'd thought about it": "Roll of Two-Time Carnegie Medal Recipients," Carnegie Hero Fund Commission, Oct. 4, 2021, www.carnegiehero.org.

80   "I always do it": Dan Majors, "Florida Man Wins Carnegie Award for Heroism a Second Time," *Pittsburgh Post-Gazette,* Dec. 17, 2014.

80   Animals may practice it: Frans B. M. de Waal, "Do Animals Feel Empathy?," *Scientific American Mind,* Dec. 2007—Jan. 2008; Mark Bekoff and Jessica Pierce, *Wild Justice: The Moral Lives of Animals* (Chicago: University of Chicago Press, 2009).

80   In 1959: Russell M. Church, "Emotional Reactions of Rats to the Pain of Others," *Journal of Comparative and Physiological Psychology* 52, no. 2 (1959): 132–34, doi.org/10.1037/h0043531.

81   three-year-old boy: "15 Years Ago Today: Gorilla Rescues Boy Who Fell in Ape Pit," CBS, Aug. 16, 2011.

81   The dolphin came up: Kathy Marks, "Save the Whales: How Moko the Dolphin Came to the Rescue of a Mother and Her Calf," *Independent,* March 13, 2008.

83   "We were lucky": Lea Winerman, "The Mind's Mirror," *American Psychological Association Monitor* 36, no. 9 (Oct. 2005).

84   "lack of general interest": L. R. Squire and T. D. Albright, *The History of Neuroscience in Autobiography: Giacomo Rizzolatti,* vol. 9 (2016): 355, Society for Neuroscience, www.sfn.org/-/media/SfN/Documents/TheHistoryofNeuroscience/Volume-9/HON_V9Rizzolatti.pdf.

84   It was later published: Giacomo Rizzolatti and Maddalena Fabbri-Destro, "Mirror Neurons: From Discovery to Autism," *Experimental Brain Research* 200, no. 3–4 (2010): 223–37, doi.org/10.1007/s00221-009-2002-3.

84   "one of the 'single most important'": Winerman, "Mind's Mirror," 48, www.apa.org, accessed March 18, 2024.

86   The Compassion-Empathy Circuit: Simon Baron-Cohen, *The Science of Evil: On Empathy and the Origins of Cruelty* (New York: Basic Books, 2011).

86   "moral-ethical" circuit: Ibid., 28.

88   videos of a human hand: Jean Decety, Chia-Yan Yang, and Yawei Cheng, "Physicians Down-Regulate Their Pain Empathy Response: An Event-Related Brain Potential Study," *NeuroImage* 50, no. 4 (2010): 1676–82, doi.org/10.1016/j.neuroimage.2010.01.025.

89   other prices to pay: Marco L. Loggia, Jeffrey S. Mogil, and Catherine M. Bushnell, "Empathy Hurts: Compassion for Another Increases Both Sensory and Affective Components of Pain Perception," *Pain* 136, no. 1 (May 2008): 168–76, doi.org/10.1016/j.pain.2007.07.017.

5. On the Nature of Psychopathic Killers and Mass Murderers

92   Empathy Quotient: "Empathy Quotient (EQ)," Psychology Tools, May 28, 2004, psychology-tools.com/test/empathy-quotient. The Empathy Quotient is a sixty-item questionnaire (there is also a forty-item version) designed to measure empathy in adults. It was developed by Baron-Cohen at the Autism Research Centre at the University of Cambridge.

92   stratify empathetic responses: Baron-Cohen, *Science of Evil,* 15–41.

94   exhibit less empathy: K. Jayasankara Reddy, Karishma Rajan Mennon, and Unnati
      G. Hunjan, "Neurobiological Aspects of Violent and Criminal Behavior: Deficits in
      Frontal Lobe Function and Neurotransmitters," *International Journal of Criminal
      Justice Sciences* 13, no. 1 (2018): 44–54, doi.org/10.5281/zenodo.1403384.

94   capacity to feel concern: Jean Decety and Yoshyia Moriguchi, "The Empathic Brain
      and Its Dysfunction in Psychiatric Populations: Implications for Intervention
      Across Different Clinical Conditions," *BioPsychoSocial Medicine* 1, no. 22 (2007),
      doi.org/10.1186/1751-0759-1-22.

94   distress cues: R. J. R. Blair, "A Cognitive Developmental Approach to Moral-
      ity: Investigating the Psychopath," *Cognition* 57, no. 1 (1995): 1–29, doi.
      org/10.1016/0010-0277(95)00676-p.

95   brain's plasticity: Sally J. Rogers and Geraldine Dawson, *Early Start Denver Model
      for Young Children with Autism: Promoting Language, Learning, and Engagement*
      (New York: Guilford Press, 2010).

95   personal nature: Barbara Bradley Hagerty, "A Neuroscientist Uncovers a Dark
      Secret," NPR, June 29, 2010, www.npr.org.

97   more aggressively: R. McDermott, D. Tingley, J. Cowden, G. Frazzetto, and D. D.
      Johnson, "Monoamine Oxidase A Gene (MAOA) Predicts Behavioral Aggression
      Following Provocation," *Proceedings of the National Academy of Sciences* 106, no.
      7 (2009): 2118–23.

98   "You see that?": Hagerty, "Neuroscientist."

99   Less than 6 percent: Bryan Wendell, "What Percentage of Boy Scouts Become
      Eagle Scouts?," *Scouting Magazine,* July 10, 2015.

99   Charles Joseph Whitman: "Charles Joseph Whitman (1941–1966)," *Behind the
      Tower,* n.d., behindthetower.org/charles-joseph-whitman.

102   Very significant differences: Tom A. Hummer et al., "Short-Term Violent Video
      Game Play by Adolescents Alters Prefrontal Activity During Cognitive Inhibition,"
      Media Psychology 13, no. 2 (2010): 136–54, doi.org/10.1080/15213261003799854.

102   primed to exhibit: Yang Wang et al., "Short Term Exposure to a Violent Video
      Game Induces Changes in Frontolimbic Circuitry in Adolescents," *Brain Imaging
      and Behavior* 3 (2008): 38–50, doi.org/10.1007/s11682-008-9058-8.

102   one out of every five: "Section 102: Juvenile Crime Facts," Jan. 22, 2020,
      U.S. Department of Justice Archives, www.justice.gov/archives/jm/
      criminal-resource-manual-102-juvenile-crime-facts.

103   "Thinking is man's": Ayn Rand, *Atlas Shrugged,* www.goodreads.com/
      quotes/7242315-thinking-is-man-s-only-basic-virtue-from-which-all-the, acessed
      March 17, 2024.

103   More than half: "68% of the World Population Projected to Live in Urban Areas
      by 2050, Says UN," May 16, 2018, United Nations Department of Economic and
      Social Affairs, www.un.org.

6. On the Nature of Prejudice, Persecution, and Racism

106   "In my work": Gustave M. Gilbert, *Nuremberg Diary* (New York: Farrar, Straus,
      1947), cited in Chikako Ozawa–de Silva, *The Anatomy of Loneliness: Suicide, Social
      Connection, and the Search for Relational Meaning in Contemporary Japan* (Berke-
      ley: University of California Press, 2021), 19.

107   genocidal dehumanization: Gail B. Murrow and Richard Murrow, "A Hypothetical
      Neurological Association Between Dehumanization and Human Rights Abuses,"
      Journal of Law and the Biosciences 2, no. 2 (2015): 336–64.

107   "any kind of communication": "United Nations Strategy and Plan of Action

on Hate Speech," Secretary-General of the United Nations, May 2019, tinyurl. com/323cnx2m.

107   It requires our volition: Murrow and Murrow, "Hypothetical Neurological Association."

108   "subhuman": Emile Bruneau et al., "Denying Humanity: The Distinct Neural Correlates of Blatant Dehumanization," *Journal of Experimental Psychology: General* 147, no. 7 (2018): 1078–93, doi.org/10.1037/xge0000417.

110   "The function": Toni Morrison, "A Humanist View" (speech, Portland State University, May 30, 1975), Oregon Public Speakers Collection, Portland State University Library, tinyurl.com/dksn5934.

110   members of our own race: Siyang Luo et al., "Oxytocin Receptor Gene and Racial Ingroup Bias in Empathy-Related Brain Activity," *NeuroImage* 110 (2015): 22–31.

112   Individuals who are: Mario F. Mendez, "A Neurology of the Conservative-Liberal Dimension of Political Ideology," *Journal of Neuropsychiatry and Clinical Neurosciences* 29, no. 2 (2017): 86–94, doi.org/10.1176/appi.neuropsych.16030051.

113   when we perceive: Nour Kteily, Gordon Hodson, and Emile Bruneau, "They See Us as Less Than Human: Metadehumanization Predicts Intergroup Conflict via Reciprocal Dehumanization," *Journal of Personality and Social Psychology* 110, no. 3 (March 2016): 343–70, doi.org/10.1037/pspa0000044.

114   About one thousand: Michael Marshall, "US Police Kill Up to 6 Times More Black People Than White People," *New Scientist,* June 24, 2020.

114   twice as likely: Lynne Peeples, "What the Data Say About Police Brutality and Racial Bias—and Which Reforms Might Work," *Nature,* June 19, 2020.

115   the less likely our brains: Susan T. Fiske, "From Dehumanization and Objectification, to Rehumanization: Neuroimaging Studies on Building Blocks of Empathy," *Annals of the New York Academy of Sciences* 1167 (June 2009): 31–34.

115   how much activity: Jennifer T. Kubota, Mahzarin R. Banaji, and Elizabeth A. Phelps, "The Neuroscience of Race," *Nature Neuroscience* 15, no. 7 (2012): 940–48, doi.org/10.1038/nn.3136.

115   "all look the same": Tobias Brosch, Eyal Bar-David, and Elizabeth A. Phelps, "Implicit Race Bias Decreases the Similarity of Neural Representations of Black and White Faces," *Psychological Science* 24, no. 2 (2013), doi.org/10.1177/0956797612451465.

115   Increased amygdala activity: Jennifer L. Eberhardt, "Imaging Race," *American Psychologist* 60, no. 2 (2005): 181–89.

115   brain wiring: Sarah E. Gaither, Kristin Pauker, and Scott P. Johnson, "Biracial and Monoracial Own-Race Face Perception: An Eye Tracking Study," *Developmental Science* 15, no. 6 (2012): 775–82.

116   marked preferences: Sandy Sangrigoli and Scania De Schonen, "Recognition of Own-Race and Other-Race Faces by Three-Month-Old Infants," *Journal of Child Psychology and Psychiatry* 45, no. 7 (2004): 11219–27.

116   these same babies: David J. Kelly et al., "Cross-Race Preferences for Same-Race Faces Extend Beyond the African Versus Caucasian Contrast in 3-Month-Old Infants," *Infancy* 11 (2007): 87–95.

116   racially heterogeneous: Gaither, Pauker, and Johnson, "Biracial and Monoracial."

116   "On the surface": Zara Abrams, "What Works to Reduce Police Brutality," *Monitor on Psychology* 51, no. 7 (2020): 30–38.

117   quicker to detect: Jennifer L. Eberhardt et al., "Seeing Black: Race, Crime, and Visual Processing," *Journal of Personality and Social Psychology* 87, no. 6 (2004): 876–93, doi.org/10.1037/0022-3514.87.6.876.

117   The data: B. Keith Payne, "Prejudice and Perception: The Role of Automatic and Controlled Processes in Misperceiving a Weapon," *Journal of Personality and Social*

Psychology 81, no. 2 (2001): 1–12.

117   association between black individuals: B. Keith Payne, Alan J. Lambert, and Larry
       L. Jacoby, "Best Laid Plans: Effects of Goals on Accessibility Bias and Cognitive
       Control in Race-Based Misperceptions of Weapons," *Journal of Experimental
       Social Psychology* 38, no. 4 (2002): 384–96.

117   "that bidirectional associations": Eberhardt et al., "Seeing Black," 877.

118   the darker: Douglas Starr, "Meet the Psychologist Exploring Unconscious Bias—
       and Its Tragic Consequences for Society," *Science,* March 26, 2020.

119   "overcome the tendency": Joshua Correll et al., "The Police Officer's Dilemma:
       A Decade of Research on Racial Bias in the Decision Not to Shoot," *Social and
       Personality Psychology Compass* 8, no. 5 (2014): 201–13.

7. On the Nature of Sadness, Depression, and Grief

121   lack of interest: Vladimir Maletic et al., "Neurobiology of Depression: An Inte-
       grated View of Key Findings," *International Journal of Clinical Practice* 61, no. 12
       (2007): 2030–40, doi.org/10.1111/j.1742-1241.2007.01602.x.

121   "kindling" phenomenon: Scott M. Monroe and Kate L. Harkness, "Life Stress, the
       'Kindling' Hypothesis, and the Recurrence of Depression: Considerations from a
       Life Stress Perspective," *Psychological Review* 112, no. 2 (2005): 417–45.

122   5 percent chance: Louise B. Andrew, "Depression and Suicide: Overview, Etiology
       of Depression and Suicidality, Epidemiology of Depression and Suicide," Med-
       scape, Oct. 17, 2021.

122   nearly forty-three thousand suicides: "Suicide Facts and Figures—Overnight
       Walk," American Foundation for Suicide Prevention, n.d., tinyurl.com/5n9x8k.pk.

122   "The so-called 'psychotically depressed' ": David Foster Wallace, *Infinite Jest* (Bos-
       ton: Little, Brown, 1996), quoted in Jaskiran Gill, "Make No Mistake About People
       Who Leap from Burning Windows," *Medium,* June 10, 2018, medium.com.

123   "The serotonin hypothesis": Philip J. Cowen and Michael Browning, "What Has
       Serotonin to Do with Depression?," *World Psychiatry* 14, no. 2 (2015): 158–60.

124   One study: Boadie W. Dunlop et al., "Functional Connectivity of the Subcallo-
       sal Cingulate Cortex and Differential Outcomes to Treatment with Cognitive-
       Behavioral Therapy or Antidepressant Medication for Major Depressive Disorder,"
       American Journal of Psychiatry 174, no. 6 (2017): 533–45, doi.org/10.1176/appi.
       ajp.2016.16050518.

124   recent brain fMRI study: Esther Landhuis, "Brain Imaging Identifies Different
       Types of Depression," *Scientific American,* Feb. 21, 2017.

126   Significant memory lapses: Maletic et al., "Neurobiology of Depression."

127   abnormal regulation: Femina P. Varghese and E. Sherwood Brown, "The
       Hypothalamic-Pituitary-Adrenal Axis in Major Depressive Disorder: A Brief
       Primer for Primary Care Physicians," *Primary Care Companion to the Journal of
       Clinical Psychiatry* 3, no. 4 (2001): 151–55.

128   autopsy studies: Maletic et al., "Neurobiology of Depression."

129   The global market: "Global Depression Drug Market Poised to Surge from USD
       14.51 Billion to USD 16.8 Billion by 2020," Market Research Store, May 10, 2016,
       tinyurl.com/43wrx35k.

129   Recent research: "Going Off Antidepressants," *Harvard Health,* March 25, 2020.

129   nearly sixteen million Americans: Benedict Carey and Robert Gebeloff, "Many
       People Taking Antidepressants Discover They Cannot Quit," *New York Times,*
       April 7, 2018.

130   "in spite of the shared": M. Katherine Shear, "Grief and Mourning Gone Awry:

Pathway and Course of Complicated Grief," *Dialogues in Clinical Neuroscience* 14, no. 2 (2012): 119–28.

131   "during this [grieving] process": Ibid.

131   "Grief is a game changer": Rhonda O'Neill, "The Difference Between Complicated Grief and Normal Grief," *HuffPost,* April 1, 2017.

131   3 to 8 percent: Sunil M. Shah et al., "The Effect of Unexpected Bereavement on Mortality in Older Couples," *American Journal of Public Health* 103, no. 6 (2013): 1140–45, doi.org/10.2105/AJPH.2012.301050.

132   brain scan studies: Peter J. Freed et al., "Neural Mechanisms of Grief Regulation," *Biological Psychiatry* 66, no. 1 (2009): 33–40, doi.org/10.1016/j.biopsych.2009.01.019.

133   One study evaluated: Anette Kersting et al., "Neural Activation Underlying Acute Grief in Women After the Loss of an Unborn Child," *American Journal of Psychiatry* 166, no. 12 (2009): 1402–10, doi.org/10.1176/appi.ajp.2009.08121875.

133   grief experienced: Anette Kersting et al., "Trauma and Grief 2–7 Years After Termination of Pregnancy Because of Fetal Anomalies—a Pilot Study," *Journal of Psychosomatic Obstetrics and Gynaecology* 26, no. 1 (2005): 9–14, doi. org/10.1080/01443610400022967.

134   entirely avoidable: David Litt, "The Coronavirus Crisis in the U.S. Is a Failure of Democracy," *Time,* May 20, 2020.

8. On the Nature of Happiness

138   This linkage: Morten L. Kringelbach and Kent C. Berridge, "The Functional Neuroanatomy of Pleasure and Happiness," *Discovery Medicine* 9, no. 49 (2010): 579–87.

138   These hot spots: Ibid.

140   showed higher activity: Bill Conklin, "The Role of the Brain in Happiness: Advances in Neuroscience Reveal Fascinating Details About How the Brain Works," *Psychology Today,* Feb. 19, 2013.

140   adaptation-level theory: Michael Kalloniatis and Charles Luu, "Light and Dark Adaptation," in *Webvision: The Organization of the Retina and Visual System,* ed. Helga Kolb, Eduardo Fernandez, and Ralph Nelson (Salt Lake City: University of Utah Health Sciences Center, 1995).

140   "people's judgments": Philip Brickman, Dan Coates, and Ronnie Janoff-Bulman, "Lottery Winners and Accident Victims: Is Happiness Relative?," *Journal of Personality and Social Psychology* 36, no. 8 (1978): 917–27.

141   "While winning $1 million": Ibid., 918.

141   no less happy. Ibid., 925.

141   despite the nostalgia: Robert E. Lane, "Does Money Buy Happiness?," *Public Interest* 57 (2023): 56–65. Lane's article was originally published in *Public Interest* in the fall of 1993.

141   Variables such as age: Frank M. Andrews and Stephen B. Withey, *Social Indicators of Well-Being: Americans' Perceptions of Life Quality* (New York: Plenum Press, 1976).

142   a happiness inventory: David A. Schkade and Daniel Kahneman, "Does Living in California Make People Happy? A Focusing Illusion in Judgments of Life Satisfaction," *Psychological Science* 9, no. 5 (1998): 340–46.

143   determined by heredity: David T. Lykken, *Happiness: What Studies on Twins Show Us About Nature, Nurture, and the Happiness Set Point* (New York: Golden Books, 1999).

143   "If the transitory variations": David T. Lykken and Auke Tellegen, "Happiness Is

a Stochastic Phenomenon," *Psychological Science* 7, no. 3 (1996): 186–89.

143   The diners' reactions: Andres O'Hara-Plotnik, "Wine Label Design Is More Important Than You Think," Eater, Oct. 12, 2016, www.eater.com.

144   how consumers made: Jennifer Aaker, Sepandar Kamvar, and Cassie Mogilner, "How Happiness Impacts Choice," Stanford Graduate School of Business, Working Paper no. 2084 (2011).

144   "Nesquik claims": Ibid., 4–5.

145   that is how much: "Getting to Know You," *Economist*, Sept. 11, 2014.

145   spending approximately: Molly Brown, "Nielsen Reports That the Average American Adult Spends 11 Hours per Day on Gadgets," GeekWire, March 13, 2015.

146   "Just wait": Charles Duhigg, "How Companies Learn Your Secrets," *New York Times*, Feb. 22, 2012.

146   based on habit: Ibid.

9. On the Nature of Laughter

149   "Analyzing humor": E. B. White and Katharine S. White, "The Preaching Humorist," *Saturday Review of Literature*, Oct. 18, 1941.

149   "We don't laugh": Christopher Bergland, "The Neuroscience of Contagious Laughter," *Psychology Today*, Sept. 29, 2017.

149   *Only five species:* Marina Davila Ross, Michael J. Owren, and Elke Zimmerman, "The Evolution of Laughter in Great Apes and Humans," *Communicative and Integrative Biology* 3, no. 2 (2010): 191–94.

150   The vocalizations: Susan Lubejko, "The Neuroscience of Laughter," NeuWrite San Diego, Aug. 23, 2018, neuwritesd.org.

150   Laughter serves: Maia Szalavitz, "It's No Joke: Why Laughter Kills Pain," *Time*, Sept. 14, 2011.

150   twenty times a day: Ben Healy, "What Makes Something Funny?," *Atlantic*, March 2018.

150   laughter was the result: Scott Weems, *Ha! The Science of When We Laugh and Why* (New York: Basic Books, 2014).

151   series of short notes: Lubejko, "Neuroscience of Laughter."

153   The latest theory: Peter McGraw and Joel Warner, "What Makes Something Funny? A Bold New Attempt at a Unified Theory of Comedy," *Slate*, March 23, 2014.

153   "Indifference is": Henri Bergson, *Laughter: An Essay on the Meaning of the Comic*, trans. Cloudesley Brereton and Fred Rothwell (1911; reprinted by Project Gutenberg).

153   something is humorous: McGraw and Warner, "What Makes Something Funny?"

156   "Common sense": Moshe Schein, *The Little Book of Aphorisms and Quotations for the Surgeon* (Shrewsbury: TFM, 2020), ix.

157   "I made the joyous": Norman Cousins, *Anatomy of an Illness as Perceived by the Patient: Reflections on Healing and Regeneration* (New York: W. W. Norton, 1979).

157   In six months: Sebastian Gendry, "Norman Cousins Anatomy of an Illness," video, Laughter Online University, May 15, 2019, www.laughteronlineuniversity.com.

157   "not just superb": Lee S. Berk et al., "Modulation of Neuroimmune Parameters During the Eustress of Humor-Associated Mirthful Laughter," *Alternative Therapies* 7, no. 2 (2001): 74.

158   one short bout: Ibid., 62–76.

158   suffered fewer episodes: S. A. Tan et al., "Mirthful Laughter: An Effective Adjunct in Cardiac Rehabilitation," *Canadian Journal of Cardiology* 13, supp. B (1997): 190.

158 from activation: Lee S. Berk et al., "Neuroendocrine and Stress Hormone Changes During Mirthful Laughter," *American Journal of Medical Sciences* 298 (1989): 390–96.

159 Only the funny videos: Szalavitz, "It's No Joke."

10. On the Nature of Stress, Fear, and PTSD

160 a rat's ability: David M. Diamond et al., "The Temporal Dynamics Model of Emotional Memory Processing: A Synthesis on the Neurobiological Basis of Stress-Induced Amnesia, Flashbulb and Traumatic Memories, and the Yerkes-Dodson Law," *Neural Plasticity* (2007): 1–33, doi.org/10.1155/2007/60803.

161 the bathtub curve: Robert M. Yerkes and John D. Dodson, "The Relation of Strength of Stimulus to Rapidity of Habit-Formation," *Journal of Comparative Neurology and Psychology* 18, no. 5 (1908): 459–82.

162 a site downriver: "Pilot Hailed for Hudson Miracle," BBC, Jan. 16, 2009.

163 "Everything [Sully] had done": Andrew Prince, "Experience, Training Make for Smooth River Landing," NPR, Jan. 16, 2009.

165 individuals under stress: Julie Streenhuysen, "Research Shows Why Some Soldiers Are Cool Under Fire," Reuters, Feb. 15, 2009.

165 high levels of BDNF: Julie Streenhuysen, "Cool Under Fire? It's All in Your Head," Reuters, Oct. 19, 2007.

166 Child abductions: Christopher Ingraham, "There's Never Been a Safer Time to Be a Kid in America," *Washington Post,* April 14, 2015.

167 "People exaggerate": Bruce Schneier, *Beyond Fear: Thinking Sensibly About Security in an Uncertain World* (New York: Copernicus Books, 2003), 25–26.

167 "experience simulator": Daniel Gilbert, *Stumbling on Happiness* (New York: Knopf, 2006).

167 nearly $1 trillion: N.B., "Is America Spending Too Much on Homeland Security?," *Economist,* April 30, 2011.

168 at least 1.5 million: Sarah Lazare, "Body Count Report Reveals at Least 1.3 Million Lives Lost to US-Led War on Terror," *Common Dreams,* March 26, 2015.

169 effect of stress on neurons: "Studies Show Stress Can Reshape the Brain," *Guardian,* Nov. 19, 2018.

169 reduces our brains: Gabriella Boston, "Long-Term Stress Damages the Brain," *Washington Times,* Dec. 2, 2009.

169 stem cells that normally: Christopher Bergland, "Chronic Stress Can Damage Brain Structure and Connectivity," *Psychology Today,* Feb. 12, 2014.

173 handful of individuals: Ed Yong, "Meet the Woman Without Fear," *Discover,* Dec. 16, 2010.

173 In none of these: Ibid.

173 Urbach-Wiethe disease: Claudio C. V. Staut and Thomas P. Naidich, "Urbach-Wiethe Disease (Lipoid Proteinosis)," *Pediatric Neurosurgery* 28, no. 4 (1998): 212–14.

173 calcium deposits: Alix Spiegel and Lulu Miller, "World with No Fear," *Invisibilia,* NPR, Jan. 15, 2015.

174 "the Doris Day test": Jennifer Altman, "The Woman Who Knows No Fear," *New Scientist,* Dec. 17, 1994.

174 exactly where happy emotions: Ibid.

174 Children under the age of ten: James D. Bremner and Meena Narayan, "The Effects of Stress on Memory and the Hippocampus Throughout the Life Cycle: Implications for Childhood Development and Aging," *Development and Psychopathology*

10, no. 4 (1998): 871–85, doi.org/10.1017/s0954579498001916.

175   Studies of blood samples: Elbert Geuze et al., "Glucocorticoid Receptor Number Predicts Increase in Amygdala Activity After Severe Stress," *Psychoneuroendocrinology* 37, no. 11 (2012): 1837–44, doi.org/10.1016/j.psyneuen.2012.03.017.

175   Having higher concentrations: Catharine Paddock, "Do Genes Play a Role in PTSD? Study of Rwanda Genocide Survivors Suggests Yes," *SOTT (Signs of the Times),* Feb. 26, 2010.

175   Administration of compounds: "PTSD Treatment Works," National Center for PTSD, U.S. Department of Veterans Affairs, www.ptsd.va.gov/professional/newsletters/research-quarterly/V15N4.pdf, accessed Apr. 20, 2022.

11. On the Nature of Music

177   wind, string, and percussion: Daniel Weinstein et al., "Group Music Performance Causes Elevated Pain Thresholds and Social Bonding in Small and Large Groups of Singers," *Evolution and Human Behavior* 37, no. 2 (March 2016): 152–58.

178   "must be ranked": Karen Schrock, "Why Music Moves Us," *Scientific American,* July 1, 2009.

178   precise synchronization: Pauline Anderson, "Cardiac Rhythms Synchronize with Music," Medscape, June 30, 2009.

179   more functional areas: "'The Power of Music' to Affect the Brain," NPR, June 1, 2011.

179   watch scenes: "What Would Your Favorite Movie Sound Like Without a Score?," Utah Symphony, Sept. 9, 2017, utahsymphony.org.

180   Timbre is what lends: Julia C. Hailstone et al., "It's Not What You Play, It's How You Play It: Timbre Affects Perception of Emotion in Music," *Quarterly Journal of Experimental Psychology* 62, no. 11 (2009): 2141–55, doi.org/10.1080/17470210902765957.

182   Music can be scrutinized: Maria Schuppert et al., "Receptive Amusia: Evidence for Cross-Hemispheric Neural Networks Underlying Music Processing," *Brain* 123 (2000): 546–59, doi.org/10.1093/brain/123.3.546.

182   we can vividly recall: Sadie Dingfelder, "Name That Tune," *Monitor on Psychology* 37, no. 7 (June 2006): 52.

183   Once considered: Sophia Yan, "Beethoven: An Unlikely Hero in China," CNN, Nov. 29, 2015.

185   men can't dance: Lauren Vinopal, "Why Men Are So Bad at Dancing, According to Science," Fatherly, Dec. 2, 2022, www.fatherly.com.

185   A 2017 survey: "1.3 Million Elementary School Students Don't Have Access to Music Classes," Children's Music Workshop, www.childrensmusicworkshop.com, accessed Dec. 26, 2019.

187   "music produces": Confucius, *Book of Rites,* cited in Steven Cornelius and Mary Natvig, *Music: A Social Experience* (New York: Routledge, 2018), 2.

12. On the Nature of Memory

188   H.M. was: Jenni Ogden, "HM, the Man with No Memory," *Psychology Today,* Jan. 16, 2012.

189   He had attended: Andy Y. Wang et al., "The Life and Legacy of William Beecher Scoville," *Journal of Neurosurgery* 137, no. 3 (2021): 886–93, doi.org/10.3171/2021.10.jns211907.

189   Massachusetts General Hospital: "Residents Alumni MGH Neurosurgery," Alumni@ Neurosurgery, Massachusetts General Hospital, alumni.neurosurgery.mgh.harvard.edu, accessed June 7, 2021.

189   "Scoville was a fearless": W. Hathaway, "Henry M: The Day One Man's Memory Died," *Hartford Courant,* Dec. 22, 2002.

190   Procedural memories: Larry R. Squire, "The Legacy of Patient H.M. for Neuroscience," *Neuron* 61, no. 1 (2009): 6–9, doi.org/10.1016/j.neuron.2008.12.023.

191   "frank about her strange": Jonathan Rée, "Permanent Present Tense: *The Man with No Memory, and What He Taught the World* by Suzanne Corkin—Review," *Guardian,* June 27, 2013.

195   "An individual's life": Daniel Kahneman and Jason Riis, "Living, and Thinking About It: Two Perspectives on Life," in *The Science of Well-Being,* ed. Felicia A. Huppert, Nick Baylis, and Barry Keverne (New York: Oxford University Press, 2006), 285–306.

198   surveyed 188 judges: "How Many Innocent People Are in Prison?," Innocence Project, Dec. 13, 2011, innocenceproject.org.

199   "How fast do you": Elizabeth F. Loftus and John C. Palmer, "Reconstruction of Automobile Destruction: An Example of the Interaction Between Language and Memory," *Journal of Verbal Learning and Verbal Behavior* 13 (1974): 585–89.

199   In nearly one-third: Kenneth S. Pope, "Memory, Abuse, and Science: Questioning Claims About the False Memory Syndrome Epidemic," *American Psychologist* 51, no. 9 (1996): 957–74.

199   can our brains: Heidi Ledford, "False Memories Show Up in the Brain," *Nature,* Nov. 6, 2007, doi.org/10.1038/news.2007.220.

200   a collaborative study: Craig E. L. Stark, Yoko Okado, and Elizabeth F. Loftus, "Imaging the Reconstruction of True and False Memories Using Sensory Reactivation and the Misinformation Paradigms," *Learning and Memory* 17, no. 10 (2010): 485–88.

202   cognitive test batteries: Melinda Smith and Lawrence Robinson, "Age-Related Memory Loss," HelpGuide.org, www.helpguide.org, accessed March 18, 2024.

202   "There is a special horror": Michael Kinsley, "Have You Lost Your Mind?," *New Yorker,* April 21, 2014.

13. On the Nature of Time

207   "If the visual brain": David Eagleman, "Brain Time," Edge, June 23, 2009, www.edge.org.

208   If we add up all: "How Fast Is the Average Blink?," Soma Technology International, Feb. 17, 2022, www.somatechnology.com.

210   The medical term: "A Woman with Dyschronometria Shares Her Experience of Losing Track of Time," *All Things Considered,* NPR, May 18, 2020.

210   a few milliseconds: William J. Matthews and Warren H. Meck, "Time Perception: The Bad News and the Good," *Wiley Interdisciplinary Reviews. Cognitive Science* 5, no. 4 (2014): 429–46, doi.org/10.1002/wcs.1298.

211   professional musicians: Daniel J. Cameron and Jessica A. Grahn, "Enhanced Timing Abilities in Percussionists Generalize to Rhythms Without a Musical Beat," *Frontiers in Human Neuroscience* 8 (2014), doi.org/10.3389/fnhum.2014.01003.

211   Many think the drummer: "Which Drummers Are the Best at Keeping Time Perfectly Without a Click Track?," Quora, Jan. 31, 2014.

211   If the timing: Roeland Hancock, Kenneth R. Pugh, and Fumiko Hoeft, "Neural Noise Hypothesis of Developmental Dyslexia," *Trends in Cognitive Sciences* 21, no. 6 (2017): 434–48, doi.org/10.1016/j.tics.2017.03.008.

211   Recent research: Suzanne Hood et al., "Endogenous Dopamine Regulates the Rhythm of Expression of the Clock Protein PER2 in the Rat Dorsal Striatum via

Daily Activation of D2 Dopamine Receptors," *Journal of Neuroscience* 30, no. 42 (2010): 14046–58, doi.org/10.1523/JNEUROSCI.2128-10.2010.

212   video of an actor: Harry McGurk and John MacDonald, "Hearing Lips and Seeing Voices," *Nature* 264 (1976): 746–48, doi.org/10.1038/264746a0.

212   MRI studies showed: Elliot D. Freeman et al., "Sight and Sound Out of Synch: Fragmentation and Renormalisation of Audiovisual Integration and Subjective Timing," *Cortex* 49, no. 10 (2013): 2875–87, doi.org/10.1016/j.cortex.2013.03.006.

215   tight synchronization: "The Human Suprachiasmatic Nucleus," HHMI BioInteractive, Feb. 4, 2000, www.biointeractive.org.

215   "a day is created": Joseph Takahashi and Michael Rosbash, "Clockwork Genes: Discoveries in Biological Time," HHMI BioInteractive, Feb. 25, 2000, www.biointeractive.org.

216   "like the mellow sunset": Christopher C. Klein, "When Edison Turned Night into Day," History Channel, Sept. 4, 2018.

217   cancer: Mei Yong and Michael Nasterlack, "Shift Work and Cancer: State of Science and Practical Consequences," *Arhiv za Higijenu Rada i Toksikologilu* 63, no. 2 (2012): 153–60, doi.org/10.2478/10004-1254-63-2012-2209.

217   obesity: X. S. Wang et al., "Shift Work and Chronic Disease: The Epidemiological Evidence," *Occupational Medicine* 61, no. 2 (2011): 78–89, doi.org/10.1093/occmed/kqr001.

218   self-administering melatonin: Andrew Herxheimer and Keith J. Petrie, "Melatonin for the Prevention and Treatment of Jet Lag," *Cochrane Database of Systematic Reviews* 2 (2002), doi.org/10.1002/14651858.CD001520.

218   Heart disease: Nen-Chung Chang et al., "Reversible Left Ventricular Apical Ballooning Associated with Jet Lag in a Taiwanese Woman: A Case Report," *International Journal of Angiology* 16, no. 2 (2007): 62–65, doi.org/10.1055/s-0031-1278250.

218   An Israeli study: Gregory Katz et al., "Time Zone Change and Major Psychiatric Morbidity: The Results of a 6-Year Study in Jerusalem," *Comprehensive Psychiatry* 43, no. 1 (2002): 37–40.

218   A separate Australian study: Michael Berk et al., "Small Shifts in Diurnal Rhythms Are Associated with an Increase in Suicide: The Effect of Daylight Saving," *Sleep and Biological Rhythms* 6, no. 1 (2008): 22–25.

218   singular experiment: Alex Kumar, "Astronauts in Caves: The Story of Siffre," AlexanderKumar.com, www.alexanderkumar.com, .accessed May 2. 2022.

220   "They have problems": Alan Burdick, *Why Time Flies: A Mostly Scientific Investigation,* 2nd ed. (New York: Simon & Schuster, 2017), 70.

221   in Sardinia: Kumar, "Astronauts in Caves."

223   "fun-time-flying": Philip A. Gable and Bryan D. Poole, "Time Flies When You're Having Approach-Motivated Fun: Effects of Motivational Intensity on Time Perception," *Psychological Science* 23, no. 8 (2012): 879–86.

223   dessert images: Sarah Glynn, "Why Time Flies When You're Having Fun," *Medical News Today,* Aug. 24, 2012.

223   we measure time: Steve Taylor, "Why Does Time Seem to Pass at Different Speeds? Does Time Really Speed Up as We Get Older?," *Psychology Today,* July 3, 2011.

225   subjects' sense of time: Ibid.

225   that notion: Ephrat Livni, "Physics Explains Why Times Passes Faster as You Age," *Quartz,* Jan. 8, 2019, qz.com.

226   Eight characteristics: Mike Oppland, "8 Traits of Flow According to Mihaly Csikszentmihalyi," *Positive Psychology,* Dec. 16, 2016, positivepsychology.com.

227   A recent study: Sylvie Droit-Volet et al., "Mindfulness Meditation, Time Judgment, and Time Experience: Importance of the Time Scale Considered (Seconds

or Minutes)," *PLoS ONE* 14, no. 10 (2019), doi.org/10.1371/journal.pone.0223567.

14. On the Nature of the Mind-Body Connection

230   75 percent of Americans: David W. Moore, "Three in Four Americans Believe in Paranormal: Little Change from Similar Results in 2001," Gallup News Service, June 16, 2005.

234   professional gambler: R. E. Langer, "René Descartes," *American Mathematical Monthly* 44, no. 8 (1937): 495, doi.org/10.2307/2301226.

235   happy, optimistic gamblers: Marjys A. Maier, Moritz C. Dechamps, and Markus Pflitsch, "Intentional Observer Effects on Quantum Randomness: A Bayesian Analysis Reveals Evidence Against Micro-psychokinesis," *Frontiers in Psychology* 9 (2018): 379, doi.org/10.3389/fpsyg.2018.00379.

236   Three thousand: Jeffrey Mishlove, "Psychokinesis," in *The Roots of Consciousness*, www.williamjames.com/Science/PK.htm, accessed March 18, 2024.

236   Recent research: Maier, Dechamps, and Pflitsch, "Intentional Observer Effects," [AU: full cite needed].

237   volunteer test subjects: Carroll B. Nash, "Psychokinetic Control of Bacterial Growth," *Journal of the Society for Psychical Research* 51, no. 790 (1982): 217–21.

237   Other studies: Allan L. Smith and Leonard Laskow, "Intentional Healing in Cultured Breast Cancer Cells," *Proceedings of the Academy of Religion and Psychical Research* (2000): 96.

238   PEAR lab investigators: John P. Bisaha and Brenda J. Dunne, "Multiple Subject and Long-Distance Precognitive Remote Viewing of Geographic Locations," in *Mind at Large: IEEE Symposia on the Nature of Extrasensory Perception,* ed. Charles T. Tart, Harold E. Puthoff, and Russell Targ (New York: Praeger Special Studies, 1979).

238   remote viewing: D. J. Dunne, R. G. Jahn, and R. D. Nelson, "Precognitive Remote Perception," Technical Note 83003, Princeton Engineering Anomalies Research Laboratory, Princeton University, 1983.

238   RVD: David Marks and Richard Kammann, "Information Transmission in Remote Viewing Experiments," *Nature* 274 (1978): 680–81.

239   "Using the standards": M. D. Mumford, PhD, A. S Rose, PhD, and D. A. Goslin, PhD., "An Evaluation of Remote Viewing: Research and Applications," American Institutes for Research, irp.fas.org/program/collect/air1995.pdf, accessed May 17, 2021.

240   But it also flouts: Robert Novella, "The Physics of ESP," New England Skeptical Society, Oct. 1998, theness.com.

241   They have been replicated: Leanna J. Standish et al., "Electroencephalographic Evidence of Correlated Event-Related Signals Between the Brains of Spatially and Sensory Isolated Human Subjects," *Journal of Alternative and Complementary Medicine* 10, no. 2 (2004): 307–14, doi.org/10.1089/107555304323062293.

241   reproduce line drawings: Ganesan Venkatasubramanian et al., "Investigating Paranormal Phenomena: Functional Brain Imaging of Telepathy," *International Journal of Yoga* 1, no. 2 (2008): 66–71, doi.org/10.4103/0973-6131.43543.

241   showed greater activity: Peiying Yang et al., "Human Biofield Therapy Modulates Tumor Microenvironment and Cancer Stemness in Mouse Lung Carcinoma," *Integrative Cancer Therapies* 19 (2020), doi.org/10.1177/1534735420940398.

241   An unrelated study: Morris Freedman et al., "Mind-Matter Interactions and the Frontal Lobes of the Brain: A Novel Neurobiological Model of Psi Inhibition," *Explore* 14, no. 1 (2018): 76–85, doi.org/10.1016/j.explore.2017.08.003.

242   Damage to the cuneus: Franziska Albrecht et al., "Unraveling Corticobasal Syn-

drome and Alien Limb Syndrome with Structural Brain Imaging," *Cortex* 117 (2019): 33–40, doi.org/10.1016/j.cortex.2019.02.015.

242   famous essay: Thomas Nagel, "What Is It Like to Be a Bat?," *Philosophical Review* 83, no. 4 (Oct. 1974): 435–50.

242   TOM may also play: Charlotte Ruhl, "Theory of Mind in Psychology: People Thinking," *Simply Psychology*, Aug. 7, 2020.

243   *the frontal lobe:* Freedman et al., "Mind-Matter Interactions."

243   more susceptible: Patrick Jones, "Mindfulness Training: Can It Create Superheroes?," *Frontiers in Psychotherapy* 10 (2019), doi.org/10.3389/fpsyg.2019.00613.

245   "Before denying": Walter B. Cannon, " 'Voodoo' Death," *American Anthropologist,* n.s., 44, no. 2 (1942): 169–81.

248   "Preservation of sound": Nick Barrowman, "The Myth of the Placebo Effect," *New Atlantis* (Winter 2016): 46–59.

250   no detectable differences: Raine Sihvonen et al., "Arthroscopic Partial Meniscectomy Versus Placebo Surgery for a Degenerative Meniscus Tear: A 2-Year Follow-Up of the Randomised Controlled Trial," *Annals of the Rheumatic Diseases* 77, no. 2 (2018): 188–95, doi.org/10.1136/annrheumdis-2017-211172.

250   "A man whom": Maj-Britt Niemi, "Cure in the Mind," *Scientific American Mind* 20, no. 1 (2009): 42–49, www.jstor.org/stable/24940066.

250   The physician's dilemma: Isadore Rosenfeld, *Dr. Rosenfeld's Guide to Alternative Medicine* (New York: Random House, 1999), 11–16.

251   widely accepted: Nina Japundžić-Žigon et al., "Sudden Death: Neurogenic Causes, Prediction, and Prevention," *European Journal of Preventive Cardiology* 25, no. 1 (2018): 29–39.

253   The individual's emotional state: Thomas E. Kraynak et al., "Functional Neuroanatomy of Peripheral Inflammatory Physiology: A Meta-analysis of Human Neuroimaging Studies," *Neuroscience and Biobehavioral Reviews* 94 (Nov. 2018): 76–92.

254   patient's clinical improvement: Marina Sergeeva et al., "Response to Peripheral Immune Stimulation Within the Brain: Magnetic Resonance Imaging Perspective of Treatment Success," *Arthritis Research and Therapy* 17 (2015): 268.

254   These positive results: Linda E. Carlson et al., "Randomized-Controlled Trial of Mindfulness-Based Cancer Recovery Versus Supportive Expressive Group Therapy Among Distressed Breast Cancer Survivors (MINDSET): Long-Term Follow-Up Results," *Psycho-oncology* 25, no. 7 (2016): 750–59, doi.org/10.1002/pon.4150.

255   breast cancer: Julienne E. Bower et al., "Mindfulness Meditation for Younger Breast Cancer Survivors: A Randomized Controlled Trial," *Cancer* 121, no. 8 (2015): 1231–40, doi.org/10.1002/cncr.29194.

255   ulcerative colitis: Sharon Jedel et al., "A Randomized Controlled Trial of Mindfulness-Based Stress Reduction to Prevent Flare-Up in Patients with Inactive Ulcerative Colitis," *Digestion* 89, no. 2 (2014): 142–55, doi.org/10.1159/000356316.

255   rheumatoid arthritis: Alex J. Zautra et al., "Comparison of Cognitive Behavioral and Mindfulness Meditation Interventions on Adaptation to Rheumatoid Arthritis for Patients with and Without History of Recurrent Depression," *Journal of Consulting and Clinical Psychology* 76, no. 3 (2008): 408–21, doi.org/10.1037/0022-006X.76.3.408.

255   MMP improved: John W. Mellors et al., "Plasma Viral Load and CD4+ Lymphocytes as Prognostic Markers of HIV-1 Infection," *Annals of Internal Medicine* 126, no. 12 (1997): 946–54, doi.org/10.7326/0003-4819-126-12-199706150-00003.

255   a separate group: Cecile A. Lengacher et al., "Lymphocyte Recovery After Breast Cancer Treatment and Mindfulness-Based Stress Reduction (MBSR) Therapy," *Biological Research for Nursing* 15, no. 1 (2013): 37–47, doi.org/10.1177/1099800411419245.

255 isolation and social loneliness: J. David Creswell et al., "Mindfulness-Based Stress Reduction Training Reduces Loneliness and Pro-inflammatory Gene Expression in Older Adults: A Small Randomized Controlled Trial," *Brain, Behavior, and Immunity* 26, no. 7 (2012): 1095–101.

255 antibody responsiveness: Richard J. Davidson et al., "Alterations in Brain and Immune Function Produced by Mindfulness Meditation," *Psychosomatic Medicine* 65 (2003): 564–70.

255 One recent study: Charlotte D'Mello and Mark G. Swain, "Immune-to-Brain Communication Pathways in Inflammation-Associated Sickness and Depression," *Current Topics in Behavioral Neurosciences* 31 (2017): 73–94, doi.org/10.1007/7854_2016_37.

257 "diminished use of the senses": Richard Louv, "What Is Nature-Deficit Disorder?," Oct. 15, 2019, richardlouv.com.

257 linking NDD: Mélanie Lapalme and Michèle Déry, "Nature et gravité du trouble de l'opposition avec provocation et du trouble des conduites selon qu'ils surviennent ensemble ou séparément chez l'enfant" [Nature and severity of oppositional defiant disorder and conduct disorder as they occur together or separately in children], *Canadian Journal of Psychiatry* 54, no. 9 (2009): 605–13, doi.org/10.1177/070674370905400905.

257 access to natural areas: Louise Chawla et al., "Green Schoolyards as Havens from Stress and Resources for Resilience in Childhood and Adolescence," *Health and Place* 28 (2014): 1–13, doi.org/10.1016/j.healthplace.2014.03.001.

258 escaping into parks: Meg St-Esprit McKivigan, " 'Nature Deficit Disorder' Is Really a Thing," *New York Times,* June 23, 2020.

258 correlated positively: Don-Gak Lee et al., "Effect of Forest Users' Stress on Perceived Restorativeness, Forest Recreation Motivation, and Mental Well-Being During COVID-19 Pandemic," *International Journal of Environmental Research and Public Health* 19, no. 11 (2022): 6675, doi.org/10.3390/ijerph19116675.

258 Being able to readily: Maya Sadeh et al., "Is Health-Related Quality of Life 1-Year After Coronary Artery Bypass Graft Surgery Associated with Living in a Greener Environment?," *Environmental Research* 212, pt. C (2022), doi.org/10.1016/j.envres.2022.113364.

258 Patients in rooms: Roger S. Ulrich, "View Through a Window May Influence Recovery from Surgery," *Science* 224, no. 4647 (1984): 420–21, doi.org/10.1126/science.6143402.

258 Individuals who lived: Geoffrey H. Donovan, Demetrios Gatziolis, and Jeroen Douwes, "Relationship Between Exposure to the Natural Environment and Recovery from Hip or Knee Arthroplasty: A New Zealand Retrospective Cohort Study," *BMJ Open* 9, no. 9 (2019), doi.org/10.1136/bmjopen-2019-029522.

259 Epidemiological studies: Lin Yang et al., "Effects of Urban Green Space on Cardiovascular and Respiratory Biomarkers in Chinese Adults: Panel Study Using Digital Tracking Devices," *JMIR Cardiology* 5, no. 2 (2021), doi.org/10.2196/31316.

259 "I shall live": "The Night's Watch," *Game of Thrones,* gameofthrones.fandom.com.

15. On the Nature of Religion and Spirituality

261 "Deep is the well": Thomas Mann, *Joseph and His Brothers: The Stories of Jacob, Young Joseph, Joseph in Egypt, Joseph the Provider,* trans. John E. Woods (New York: Everyman's Library, 2005).

261 "one of the most important": Nicholas Birch, "7,000 Years Older Than Stonehenge: The Site That Stunned Archeologists," *Guardian,* April 22, 2008.

262 potential sites: Andrew Curry, "Gobekli Tepe: The World's First Temple?," *Smith-sonian Magazine,* Nov. 2008.

262 Klaus Schmidt: Patrick Symmes, "Turkey: Archeological Dig Reshaping Human History," *Newsweek,* Feb. 18, 2010.

264 impact of faith: Robert N. Bellah, *Religion in Human Evolution: From the Paleo-lithic to the Axial Age* (Cambridge, Mass.: Harvard University Press, 2011).

264 theoretical constructs: Andrew Brown, "Religion in Human Evolution, Part I: The Co-evolution of Gods and Humanity," *Guardian,* July 16, 2012.

264 Dr. Jeffrey Anderson: Michael A. Ferguson et al., "Reward, Salience, and Atten-tional Networks Are Activated by Religious Experience in Devout Mormons," *Social Neuroscience* 13, no. 1 (2018): 104–16.

266 "And in that moment": S. Bangambam, "My Stroke of Insight by Jill Bolte Taylor (Transcript)," *Sinju Post,* Sept. 1, 2014, singjupost.com.

266 "And I look down": Ibid.

267 "a growing sense": Jill Bolte Taylor, *My Stoke of Insight: A Brain Scientist's Personal Journey* (New York: Penguin, 2009), 41.

267 "two fundamentally": Pete Enns, "Religious Faith: It's More a Right Brain Thing," PeteEnns.com, April 29, 2017, peteenns.com.

268 "the tendency": Cosimo Urgesi et al., "The Spiritual Brain: Selective Cortical Lesions Modulate Human Self-Transcendence," *Neuron* 65, no. 3 (2010): 309–19, doi.org/10.1016/j.neuron.2010.01.026.

268 Research studies of twins: Katherine M. Kirk, Lindon J. Eaves, and Nicholas G. Martin, "Self-Transcendence as a Measure of Spirituality in a Sample of Older Australian Twins," *Twin Research* 2, no. 2 (1999): 81–87.

268 Recent studies show: Jacqueline Borg et al., "The Serotonin System and Spiri-tual Experiences," *American Journal of Psychiatry* 160, no. 11 (2003): 1965–69, doi.org/10.1176/appi.ajp.160.11.1965.

268 Italian researchers: Urgesi et al., "Spiritual Brain."

269 handful of patients: Stephen G. Waxman and Norman Geschwind, "The Interictal Behavior Syndrome of Temporal Lobe Epilepsy," *Archives of General Psychiatry* 32, no. 12 (1975): 1580–86, doi.org/10.1001/archpsyc.1975.01760300118011.

270 Another patient: Ibid., 1585.

270 Of these patients: Frank Tong, "Out-of-Body Experiences: From Penfield to Pres-ent," *Trends in Cognitive Sciences* 7, no. 3 (2003): 104–6.

270 as Penfield discovered: Jane E. Aspell and Olaf Blanke, "Understanding the Out-of-Body Experience from a Neuroscientific Perspective," in *Psychological Scientific Perspectives on Out-of-Body and Near-Death Experiences,* ed. Craig D. Murray (New York: Nova Science, 2009).

270 suggested that OBEs: Susan Blackmore, *Beyond the Body: An Investigation of Out-of-Body Experiences* (Chicago: Academy Chicago Publishers, 1992).

271 Olaf Blanke: Shahar Arzy et al., "Induction of an Illusory Shadow Person," *Nature* 443 (2006): 287.

271 effects of cortical stimulation: Sandra Blakeslee, "Out-of-Body Experience? Your Brain Is to Blame," *New York Times,* Oct. 3, 2006.

271 "multisensory conflict": H. Henrik Ehrsson, "The Experimental Induction of Out-of-Body Experiences," *Science* 317, no. 5841 (2007): 1048.

272 underlying cause: Amanda Onion, "What Causes Spooky Out-of-Body Experi-ences? Inner Ear Problems Could Be to Blame," *Scientific American Live Science,* Aug. 6, 2017.

16. On the Nature of the Brain of the Future

274   "Why can't I ask": Madeline Lancaster, "Growing Mini Brains to Discover What Makes Us Human," video, TEDxCERN, Nov. 30, 2015, www.youtube.com/watch?v=EjiWRINEatQ.

276   "Well, we survived": Denise Chow, "Will Humanity Survive This Century? Sir Martin Rees Predicts 'a Bumpy Ride' Ahead," Mach-NBC News, Nov. 8, 2018.

280   "In contrast to": P. B. Rutledge, "Augmented Reality: Brain-Based Persuasion Model," www.psychologytoday.com/sites/default/files/rutledge_ar_brain-based_model_of_persuasion.pdf.

280   Let's agree: Melinda Wenner Moyer, "A Safe Drug to Boost Brainpower," *Scientific American,* March 1, 2016.

281   "doping with smart drugs": Amber Dance, "Smart Drugs: A Dose of Intelligence," *Nature* 531 (2016): S2—S3, doi.org/10.1038/531S2a.

281   "nootropics challenge": Josh Helton, "Nootropics: Can a Smart Pill Increase Your Focus and Intelligence? Let's Find Out," *Hustle,* Nov. 5, 2015.

282   "At this point": Josh Helton, "Nootropics: What Happened When I Went 30 Days on Smart Drugs," *Hustle,* Dec. 8, 2015, https://thehustle.co/i-spent-four-weeks-taking-nootropics-heres-what-i-discovered, accessed Oct. 17, 2017.

282   long list: For further information, see Mitch Gilliland, *The Complete Bible of Nootropics and Cognitive Enhancers* (San Bernardino, Calif.: self-published, 2017).

283   a recent scandal: Jennifer Medina, Katie Benner, and Kate Taylor, "Actresses, Business Leaders, and Other Wealthy Parents Charged in U.S. College Entry Fraud," *New York Times,* Oct. 6, 2021.

283   machine learning system: Matt Reynolds, "AI Learns to Write Its Own Code by Stealing from Other Programs," *New Scientist,* Feb. 22, 2017.

284   They would be able: Ray Kurzweil, *The Singularity Is Near* (New York: Penguin Books, 2006).

284   "The development": Victor Luckerson, "5 Very Smart People Who Think Artificial Intelligence Could Bring the Apocalypse," *Time,* Dec. 2, 2014.

285   Musk called for: Ibid.

286   the risk of falling: Nick Bostrom, "Existential Risks: Analyzing Human Extinction Scenarios and Related Hazards," *Journal of Evolution and Technology* 9, no. 1 (2002): [AU: Page nos. needed.].

286   sobering reminder: Anthony Berglas, "Artificial Intelligence Will Kill Our Grandchildren (Singularity)," Jan. 2012, berglas.org.

286   "We know that all": Ibid.

Epilogue: Better Angels

288   "Men, it has been": Charles Mackay, *Extraordinary Popular Delusions and the Madness of Crowds* (London, 1841).

290   "had deteriorated": Ewa Pieta, "The Red Button and the Man Who Saved the World," Logtv.com, n.d.

292   "There was no rule": Greg Myre, "Stanislav Petrov, 'the Man Who Saved the World,' Dies at 77," NPR, Sept. 17, 2017.

294   "Only a human being": "The Case for the Petrov Rule," Project Ploughshares, June 7, 2021, ploughshares.ca.

295   overall probability: Anders Sandberg and Nick Bostrom, "Global Catastrophic Risks Survey," Technical Report no. 2008, Future of Humanity Institute, Oxford University.

295   "human extinction event": "Global Catastrophic Risk," Wikipedia. <~?~AU: Please

provide a more authoritative source for this information.>

297 more than thirty countries: Jonah M. Kessel, "Do I Need to Worry About a Terminator Knocking at My Door?," *New York Times,* Dec. 13, 2019.

298 "Artificial intelligence": Vasily Sychev, "The Threat of Killer Robots," *UNESCO Courier,* June 25, 2018, en.unesco.org.

298 In a recent TED talk: Steven Pinker, "Is the World Getting Better or Worse? A Look at the Numbers," TED.com, April 30, 2018.

299 In 1960, the global literacy rate: M. Roser and E. Ortiz-Ospina, "Our World in Data: Literacy," OurWorldInData.org, 2016.

299 "Natural experiments": Nicholas A. Christakis, *Blueprint: The Evolutionary Origins of a Good Society* (Boston: Little, Brown, 2019), 25.

299 "Survivor camps": Ibid, 26.

301 "Capacities of the Social Suite": Ibid., 13.

302 "We are not enemies": Abraham Lincoln, First Inaugural Address (1861), American Presidency Project, www.presidency.ucsb.edu.

303 girl trying to make her way: Kevin Carter, photo of a Sudanese child and a vulture, March 1993, en.wikipedia.org/wiki/File:Kevin-Carter-Child-Vulture-Sudan.jpg.

303 three-year-old boy: Adam Withnall, "If These Extraordinarily Powerful Images of a Dead Syrian Child . . . ," *Independent,* Sept. 3, 2015.

303 the Fermi paradox: "New Study Estimates the Odds of Life and Intelligence Emerging Beyond Our Planet," *ScienceDaily,* May 18, 2020, www.sciencedaily.com.

304 "The view of the Earth": *Congressional Record, Hearings, Reports, and Prints of the House Committee on Appropriations: Departments of Labor and Health, Education, and Welfare Appropriations for 1970* (Washington, D.C.: U.S. Government Printing Office, 1970), 1002.

305 "The isolation of Earth": "Apollo's Daring Mission," *Nova,* PBS, Dec. 26, 2018.

305 "The eternal silence": Blaise Pascal, *The Thoughts of Blaise Pascal* (London: Kegan Paul, 1885).

Index

List of Illustration Sources and Credits

001 "Dr. Jose Delgado, 1952–1967," folder 836, box 54, Yale University Manuscripts and Archives. Courtesy of Yale University.

002 Diana Reiss and Lori Marino, "Mirror Self-Recognition in the Bottlenose Dolphin: A Case of Cognitive Convergence," *Proceedings of the National Academy of Sciences* 98, no. 10 (May 2001): 5937–42. Courtesy of *Proceedings of the National Academy of Sciences.*

003 Jean-Jacques Perrissin, *The Fatal Joust of King Henry II of France on 30 June 1549,* 1570.

004 M. Gray, *Calcarine sulcus. Medial surface of left cerebral hemisphere,* fig. 727 from the 1918 edition of *Gray's Anatomy.*

005 Photo by Charles F. Badland, 2006, Creative Commons Attribution-Share Alike 2.0 Generic License.

006 Photo by ? Creative Commons Attribution-Share Alike 4.0 International License.

007 Photo by the author.

008 Photo by U.S. Embassy Jerusalem, Creative Commons Attribution 2.0 Generic.

009 Harry T. Chugani et al., "Local Brain Functional Activity Following Early Deprivation: A Study of Postinstitutionalized Romanian Orphans," *NeuroImage* 14, no. 6 (2001): 1290–301. Courtesy of *NeuroImage.*

010 Photo by Mstyslav Chernov, 2015, Creative Commons Attribution-Share Alike International License.

011 Vittorio Gallese, Luciano Fadiga, Leonardo Fogassi, and Giacomo Rizzolatti, "Action Recognition in the Premotor Cortex," *Brain* 132 (2009): 1686.

012 Jaime A. Pineda, "Sensorimotor Cortex as a Critical Component of an 'Extended' Mirror Neuron System: Does It Solve the Development, Correspondence, and Control Problems in Mirroring?," *Behavioral and Brain Functions,* Oct. 18, 2008. Courtesy of Jaime A. Pineda, PhD.

014 J. L. Eberhardt, P. A. Goff, V. J. Purdie, and P. G. Davies, "Seeing Black: Race, Crime, and Visual Processing," *Journal of Personality and Social Psychology* 87, no. 6 (2004): 879. Courtesy of Jennifer L. Eberhardt, PhD.

015 Lise Alschuler, "Hypothalamic Pituitary Adrenal (HPA) Axis," Integrative Therapeutics, Dec. 3, 2019, www.integrativepro.com.

016 Author's own illustration.

017 A model of the evolution of laughter among the great apes and humans. Permission to use figure courtesy of Dr. Marina Davila-Ross

017 M. Davila-Ross, M. Owren, M., and E. Zimmermann, E. (2010). "The Evolution

of Laughter in Great Apes and Humans," *Communicative & Integrative Biolog* 3, no. 2 (2010): 191–94, doi.org/10.4161/cib.3.2.10944.

018   David M. Diamond et al., "The Temporal Dynamics Model of Emotional Memory Processing: A Synthesis on the Neurobiological Basis of Stress-Induced Amnesia, Flashbulb and Traumatic Memories, and the Yerkes-Dodson Law," *Neural Plasticity* (2007): 1–33. Creative Commons CCO 1.0 Universal Public Domain.

019   Author's own illustration.

020   The Treasures," Marine Science, 2003, marinebio.net. Courtesy of Genny Anderson, Creative Commons Attribution-Share Alike 4.0 International License.

021   This image is a composite assembled from data acquired by the Suomi NPP satellite in April and October 2012. NASA Earth Observatory/NOAA NGDC.

022   Tamar Shochat and Eran Tauber, "Daily Rhythms of the Body and the Biological Clock," *Frontiers for Young Minds* 9 (2021). Courtesy of Tamar Shochat and Eran Tauber, Creative Commons Attribution License.

023   Stephen Pankhurst, "Why Does Time Seem to Go by Faster as One Gets Older? Is This a Completely Subjective Experience That Varies?," *Quora,* Jan. 4, 2015. Courtesy of Stephen Pankhurst.

024   Based on work by Wilder Penfield. "Anatomy and Physiology," Connexions, cnx.org. Creative Commons Attribution 3.0 Unported License.

025   Stephan Schwartz, "Remote Viewing," *Psi Encyclopedia* (London: Society for Psychical Research, 2017), psi-encyclopedia.spr.ac.uk.

A NOTE ON THE TYPE

This book was set in Minion, a typeface produced by the Adobe Corporation specifically for the Macintosh personal computer and released in 1990. Designed by Robert Slimbach, Minion combines the classic characteristics of old-style faces with the full complement of weights required for modern typesetting.

Composed by North Market Street Graphics,
Lancaster, Pennsylvania

Designed by Cassandra J. Pappas